Biogenetic Structuralism

Biogenetic Structuralism

Charles D. Laughlin Jr.
&
Eugene G. d'Aquili

1974
Columbia University Press
New York and London

Library of Congress Cataloging in Publication Data

Laughlin, Charles D 1938-
Biogenetic structuralism.

Includes bibliographies.
1. Neuropsychology. 2. Neurobiology. 3. Structural anthropology. I. d'Aquili, Eugene G., 1940- joint author. II Title. [DNLM: 1. Philosophy. B841.4 L374b]
QP356.L37 612'.8 74-13245
ISBN 0-231-03817-8

Printed in the United States of America

This Book is Dedicated to
Our Children

Beth, Sarah, and Kate
Julie and Ellie

Contents

Acknowledgments

A NUMBER OF FRIENDS and colleagues have read earlier drafts of this book. We wish to express our grateful appreciation for their helpful comments and criticism. They are: Professors Aaron T. Beck, William Chambers, Ward Goodenough, Solomon Katz, Allan Mann, John McManus, A. F. C. Wallace, and Mr. Robert Rubinstein.

As always with a book of this nature many persons influenced us either directly or indirectly in its production. Often this was accomplished by their reading previous papers upon which present chapters are based, or else in personal conversations with us. We wish to thank the following persons: Professor Charles Ackerman, Professor John Cove, the late Professor Walter Cywinski, Professor Lynn Hoffman, Professor B. J. Isbell, Dr. Jeffrey Keefe, Ms. Patricia Kolarik, Ms. Elizabeth Laughlin, Ms. Martha Laughlin, Professor Jerre Levy, Professor James McCroskery, Mr. Terry Morse, Professor Robert Pankin, Mr. A. J. Pfeiffer, Professor Iain Prattis, Professor David M. H. Richmond, Mr. Leigh Sherman, Professor W. John Smith, Professor Albert Stunkard, and Professor Howard Walker.

Last, but certainly not least, we wish to thank Ms. Mary Lovell, Ms. Kathy Casella, Ms. Mary Lou d'Aquili, and Ms. Helen Raby, who gave us so much assistance in the preparation of the manuscript.

Acknowledgments

Biogenetic Structuralism

When studied narrowly in himself by anthropologists or jurists, man is a tiny, even a shrinking, creature. His overpronounced individuality conceals from our eyes the whole to which he belongs; as we look at him our minds incline to break nature up into pieces and to forget both its deep inter-relations and its measureless horizons: we incline to all that is bad in anthropocentrism. And it is this that still leads scientists to refuse to consider man as an object of scientific scrutiny except through his body.—

TEILHARD DE CHARDIN, *The Phenomenon of Man*

I

Introduction to Structuralism

> As a first approximation, we may say that a structure is a system of transformations. Inasmuch as it is a system and not a mere collection of elements and their properties, these transformations involve laws: the structure is preserved or enriched by the interplay of its transformation laws, which never yield results external to the system nor employ elements that are external to it. In short, the notion of structure is comprised of three key ideas: the idea of wholeness, the idea of transformation, and the idea of self-regulation.—JEAN PIAGET, *Structuralism*

LIKE EXISTENTIALISM in the 1950's, structuralism seems to be the topic of the decade. Unfortunately, and also like existentialism, few people appear to be able to define the movement although almost everyone has heard of it. Part of this confusion arises from the fact that "structural approaches" have arisen in several disciplines simultaneously, notably in anthropology, political science, mathematics, linguistics, and to some extent in neurophysiology. Since some of the meaning of structuralism varies with the discipline in which it is conceived, the intelligent reader may be pardoned his confusion in the presence of ambiguity.

Philosophical Status of Structuralism

We are not concerned here with tracing the history of the various permutations of structuralist thought throughout the ages. Our primary concern is the definition of what we feel to be a new synthesis of structuralist thinking arising from modern theoretical anthropology, but with wide-ranging implications throughout the behavioral and social sciences generally. There is little doubt that the French school of structural anthropology as currently expressed in the work of Claude Levi-Strauss and his followers is most closely linked with our field of concern. A number of controversial and crucial issues have been raised by orthodox structuralists that are reflected to a greater or lesser degree in the other branches of science. These issues are demanding attention from educated persons in general, and the demand is felt for a comprehensive answer to the question, What is structuralism?

Strong, almost metatheoretical statements are being advanced in several quarters that leave the uninitiated wondering if science is indeed on the verge of stumbling over the philosopher's stone. Thus Stent (1972), in somewhat rhetorical fashion, states:,

> It is only in the past 20 years or so, more or less contemporaneously with the growth of molecular biology, that a resolution of the age-old epistemological conflict of materialism vs. idealism was found in the form of what has come to be known as structuralism.
>
> . . . Both materialism and idealism take it for granted that all the information gathered by our senses actually reaches our mind; materialism envisions that thanks to this information reality is mirrored in the mind, whereas idealism envisions that thanks to this information reality is constructed by the mind. Structuralism, on the other hand, has provided the insight that knowledge about the world enters the mind not as raw data but in some already highly abstracted form, namely as structures. In the preconscious process of converting the primary data of our experience step by step into structure, information is necessarily lost, because the creation of structures, or the recognition of patterns, is nothing less than the selective destruction of information. Thus since the mind does not gain access to the full set of data about the world, it can neither mirror nor construct reality. Instead for the mind reality is a set of structural transforms of primary data taken from the world. This transformation process is hierarchical, in that "stronger" structures are formed from "weaker" structures through selective destruction of information. Any set of primary data becomes meaningful only after a series of such operations has so transformed it that it has become congruent with a stronger structure preexisting in the mind. Neurophysiological studies carried out in recent years on the process of visual per-

> ception in higher mammals have not only shown directly that the brain operates according to the tenets of structuralism but also offer an easily understood illustration of those tenets. (pp. 92-93)

We, who propose and embrace a special brand of structuralism, i.e., biogenetic structuralism, have very warm feelings for this brief exposition of the philosophical tenets of structuralism. Stent has hit upon some of the most critical aspects of the approach, but more importantly has seen, as few have done before, the necessary bond between structural analysis of human society and the study of the human central nervous system. Furthermore, he has not perpetuated the logical error of assigning structuralism to the epistemological status of a method or construed it as a philosophical movement intermediate between materialism and idealism. Structuralism, as we view the movement, is fully a philosophy and a body of empirically based theory in its own right. Whereas materialism (say as represented by the philosophy of John Locke or David Hume) asserts that the material conditions of reality impose their intrinsic order upon the mind, and idealism (including the philosophies of Kant and Hegel, as well as phenomenology or existentialism) asserts that meaning and order are attributed to the world by the mind, structuralism (which includes the work of Karl Marx) holds that order resides both in the world and in the mind, and that the order at both loci are somehow isomorphic.[1]

Stent, in his enthusiasm, unfortunately has offered two misrepresentations that are as instructive for our purpose as his representations. In the first place, although his is a fairly good statement of what structuralism *ought to be*, it is most certainly not what is generally presented as structuralism. Structuralism, even biogenetic structuralism, does not resolve once and for all the idealist-materialist controversy. And though his formulation may explain the indeterminacy and mutability of knowledge when one starts with a basically materialistic proclivity, we seriously doubt it would convince a doc-

[1] It will be easier perhaps for the interested reader to understand the different conceptions of the nature of the dialectic in the controversy between Levi-Strauss and Jean Paul Sartre (see Levi-Strauss, 1966a, final chapter) if he keeps these philosophical strains in mind. Whereas Levi-Strauss reflects a structuralist notion of dialectic process, Sartre exhibits more a phenomenological position.

trinaire Hegelian (assuming, of course, there are any left around). Second, Stent gives the uninitiated reader the impression that the impetus to "think structurally" somehow dates from certain crucial biological discoveries. While it is true that the discovery of the molecular structure of DNA and RNA, and their long-range effect on genetic coding, cast a definite structural tint on molecular biology, they did not provide the theoretical undergirding for the structural view of human behavior in general—a view that has come to preoccupy so many contemporary scientists. Furthermore, although certain neuroanatomical and neurophysiological discoveries have indicated clear structural-functional relationships in the brain, such as the organization of cells in the calcarine cortex for the perception of lines, angles, curves, etc., these findings did not generate the pervasive, indeed metatheoretical interest in structuralism as a movement. As a point of fact it has been the worldwide interest in the work of Levi-Strauss, and later Noam Chomsky, that seems to have made structuralism a watchword for many.

Themes in Orthodox Structuralism

In order to define what we mean by biogenetic structuralism, it is necessary to develop a more comprehensive understanding of orthodox structuralism. And in order to characterize structuralism we must examine the definitive tenets or themes that have been combined to form the theory of structures itself. Like many another complex mode of thought, the conception of structure is comprised of a number of notions that have been "in the air" for a long time. For present purposes, the concept of structure may be seen as the compilation of at least three themes—systems, surface and deep levels of reality, and the unconscious.

STRUCTURE AS SYSTEM

As generally formulated, structures are viewed as naive systems. That is, structures are comprised of elements of some sort and the rules of their combination. Structures thus form configurations, the meaning or total impact of which cannot be understood apart from the set of relationships between elements. This is really a restatement of philosophical holism present in Bergson (1907), Whitehead (1929), and later perfected in the general system theory of von Bertalanffy (1956-1971, cg. 1968). In this immediate sense the structuralist-functionalist controversy that was waged in anthropology during the first half of this century was also a very lively topic in ancient Greece—Plato's *Timaeus* certainly may be considered a structuralist document.

But more properly, modern structuralists, at least in the social sciences, trace their lineage back to the work of Emile Durkheim, a social philosopher who more than any other epitomized the application of naive systems theory to an analysis of human society. Durkheim (reprint 1972) was intensely concerned with the distinction he saw between what he called "mechanical" and "organic" solidarity in groups. Mechanical solidarity was the mode of unification characteristic of the most primitive and homogeneous societies. Under the pressure of demographic expansion, however, internecine feuds became rife in any homogeneous group, and this resulted either in the destruction of the group, or in the modification of the group structure to systematically inculcate two or more roles filled alternately with mutually dependent persons (e.g., the sexual division of labor). The fission of society into various parts creates the problem of reunification at each subsequent level of solidarity. The mutual dependence of parts or elements and their reunification into an integrated whole is the *sine qua non* of *organic* solidarity, the term organism being a common term for "system" at the time Durkheim wrote.

For Durkheim, as for A. L. Kroeber with his conception of culture, social structure was seen as a "superorganic" phenomenon. He did not consider individual psychology, much less neurophysiology, as a locus for explanations of society. He contended that social phenomena can and should be explained only with reference to other social "facts." To do otherwise was, for Durkheim, an unforgivable reductionism. Of course, some of his students did not follow his example on this point. Robert Hertz (1909), a particularly gifted and brilliant student of society, resorted to both sociological and neurophysiological formulations in his explanation of right-handedness in man. Another of Durkheim's students, and one of the few to survive the First World War, Marcel Mauss—and later Mauss's student Claude Levi-Strauss—carried social analysis to the psychological-cognitive level. They did this rather self-consciously and gave some lip service to their master's anti-reductionist pronouncements. In *The Gift* (1925), Mauss analyzes the social structure of exchange and gift giving by at least alluding to psychological motives and consequences, although these are clearly projected upon the social level and institutionalized. It is upon this development (combined with influence from the Prague school of structural linguistics) that Levi-Strauss constructed the grand structural edifice that is widely understood to represent structuralism.

Levi-Strauss conceives of structure as underlying in some fashion the patterns of social behavior and cognition characteristic of particular societies. But unlike behavior (including institutionalized and ritualized behavior) that may appear to vary widely in the descriptions of ethnographers, structures are finite in number and universal among men. This is because human behavior is the product of human reason, and the operations of human reason are similar for all men everywhere.

As with all structuralists, Levi-Strauss' structures are systems. That is, they are comprised of a finite number of elements that may be combined and recombined according to a set of rules (i.e., transformational laws). It is thus through the reshuffling of elements inherent in the structure of all primitive social interaction that Levi-Strauss shows the essential equivalence, at the structural level, of systems of totemic relations in Australia on the one hand and Indian caste systems on the other (1966a).

We do not intend at the outset to portray Levi-Strauss' conception of structure in all its rich detail; we will return to this particular topic in later chapters. Yet the structural approach has engendered a good bit of confusion, and it might be helpful to depict a simple example of what is meant by a structural analysis. A Levi-Straussian structuralist will examine a real body of data (perhaps a society or set of societies) and will single out an institution Q for analysis. "Q" may be a myth, "divine kingship," kin terminological system, kinship relations—in fact almost anything. He then extracts the fewest number of essential, irreducible elements that comprise Q. Normally there will be at least two such elements (or sets of elements). Let us call the elements α and β. Careful analysis indicates that precisely because α and β are *essential*, they are therefore sociocultural markers (symbols) and usually represent extremes along some conceptual dimension. Thus α may be "high" along the dimension of height and β may be "hot" along the dimension of temperature.

At this point the structuralist may introduce a statement about primitive (we should say *basic*) mental functioning on the basis of considerable ethnographic data. He might state that the opposite of the above elements (high-hot) are present to the mind automatically and necessarily in the same way that the phenomenologists would assert that the obverse of a coin is understood on some level (by "apresentation") when the reverse is referred to.[2] Let us then label these contrasting opposites as γ "low," and δ "cold." If we have indeed adequately described the essential elements comprising the institution Q to begin with, and the relationship of these elements to the contrasting elements derived by deducing the equally essential dimensions, then we can construct the structure underlying Q as presented in Figure 1.1.

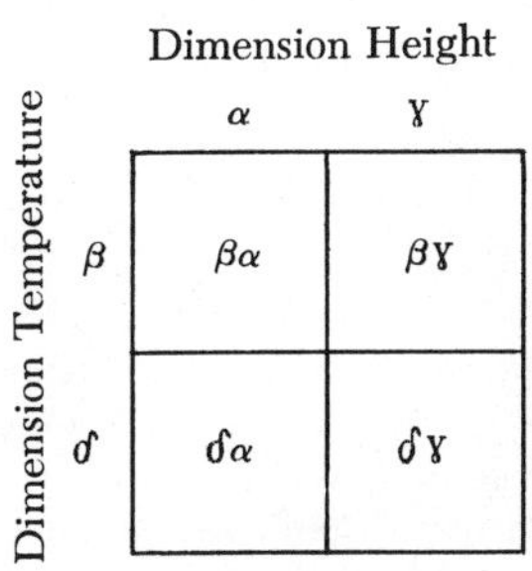

Figure 1.1. Structure of Q.

[2] We cannot review here the vast body of evidence that Levi-Strauss has used to adduce the fundamental tendency in man to think in binary opposition. Instead, the reader is directed to the works of Levi-Strauss (1966a, 1967, 1969a, 1969b) himself.

From Figure 1.1 it is obvious that only cell ($\beta\alpha$) represents the original institution Q with which our structuralist began.[3] The other three cells represent possible transforms of Q within the underlying structure. Presumably the other cells represent institutions that are actually discoverable in another social situation, serving much the same function and existing in much the same context as Q. We repeat that the above example is oversimplified. The number of essential elements, and thus the quantum of permutations, may go considerably beyond two, making an extremely intricate and complex structure—not to mention the complexity of the transformational rules by means of which one gets from one cell to another. The difficulties inherent in this necessarily deductive type of social analysis will be discussed in the next section and in chapter 6.

DEEP STRUCTURE

Unlike Durkheim who, as we have seen, conceived of crucial structure existing at the level of institutionalized social behavior (so-called surface structure), the majority of structuralists today, including Levi-Strauss and Jean Piaget, see crucial structure as lying at a level below that of social behavior—at the level of "mind" or "reason" ("deep" structure). This is also not a new idea. The famous linguist, Ferdinand de Saussure (1959), pointed out the distinction between the "deep" structure of language (he termed this *la langue*), including phonemics and grammar, and the individual speech event (he termed *la parole*), which lies, as it were, on the

[3] Upon inspection, Figure 1.1 appears remarkably similar to the grid that would be generated by componential analysis as elaborated by Goodenough (1956) and Lounsbury (1964) in their study of the semantics of folk terminology. In particular their use of contrast sets is strikingly similar to that of Levi-Strauss. But there the similarity ends. Componential analysis, which has become both popular and extremely useful in anthropology, is primarily a methodology based on a calculus of the class-product form. It is an analytical tool that makes no statement about the way the human mind operates at all times and in all places. Although many investigators using componential analysis implicitly claim that any given model it generates may represent the process by which a native group thinks, this claim is not essential to using the tool, nor is it claimed to represent universal mental laws. Certainly componential analysis does not provide any system of transformations. It is therefore essentially static.

"surface" of language. All of the grammatical relations and phonemic possibilities contained within any particular language, de Saussure taught, are never manifest in each and every occurrence of a speech event. They also may not be present in the entire speech repertoire of any particular individual speaker. Thus the speech of an individual speaker will represent only a partial or imperfect apperception of the structure of his native language.

By analogy from the insight into levels of structure suggested by de Saussure and his followers, the dichotomy of structural reality into two levels has been generalized to all sociolinguistic phenomena. Furthermore, structuralists generally take the position (either implied or explicit) that structures at the deep level are in some way more "real" than structures at the surface or behavioral level. The reasoning upon which this ontological assumption is based might go something like this: Surface structures are illusory in that they will vary from individual to individual, or society to society, in space and from one occasion to another occasion in time. This will be due to a myriad of extrinsic variables that interact with each event to give a unique quality to that event. Thus how one observed the pattern of conscious belief and practice relative to marriage in Nuer in the year 1929 will be the result of such variables as climate, population, warfare, available brideprice, and personalities of politically powerful elders. More than likely these patterns will not be the same as those observed among "the same group" (the aggregate of individuals forming the group in 1929 will have changed) in the year 1970. Structures at the surface (which Levi-Strauss has termed "statistical models") then will be in a constant state of flux, and an analysis of a structure at this level will be valid only for that social group at one locus in space-time. Hence, analyses relying solely upon surface phenomena give the erroneous impression that human society exists in a state of amorphous change and that no formulations may be generated toward a global explanation of social forms. Structures at the deep level (Levi-Strauss calls these "mechanical models"), on the other hand, are not amenable to modification by extrinsic variables. An analysis of deep structure need only concern itself with variables intrinsic to the structure itself. The deep structure is an unchanging and total system that is operative at all places and at all times. Such structures are therefore more "real" and more appropriate to the purview of the science of man.

STRUCTURE AS UNCONSCIOUS

By combining the systematic nature of structure with the conception of deep structure, we have further delineated the principal concern that characterizes the structuralist movement and contrasts it sharply with other structural movements in the history of social theory [e.g.. functionalism (á la Malinowski), structural-functionalism (á la Radcliffe-Brown or Talcott Parsons)]. That leaves us with only one more theme in our brief characterization of the central tenets of the movement—the unconscious nature of structures.

In most structural analyses, it is either assumed or stated that the structural process—that is, the transformation of one structural configuration into another structural configuration—occurs at an unconscious level in the mind. A Bororo mythmaker is not aware of the fact that he has performed a transformation through the basic structure of his myth. He is aware only of telling the myth. He may also be aware of the "reasons" for telling the myth, e.g., to pass on the story, to educate the young, to entertain, to fulfill the requisite of his role. But, as we have seen, these reasons are irrelevant to a structural analysis; what is important is the myth itself.

Recourse to the unconscious in the analysis of structures stems from two sources of influence—de Saussurian structural linguistics and Freudian psychoanalysis. To return again to de Saussure, an individual speaker was assumed to be unaware of *la langue,* or even the portions of *la langue* as imperfectly or incompletely manifest in the grammar and phonemics of his own linguistic infrastructure. Rather, the naive speaker was characterized as being aware only of the ideas he is desirous of communicating and the social event within which *la parole* occurs.

Of even greater influence on structuralists (especially nonlinguistic structuralists) has been the psychoanalytic movement originated by Sigmund Freud. Freud pointed out that much of our behavior is activated by motivations of which we are scarcely aware. It is only after the fact that we construct conscious "reasons" for our behavior, so-called rationalizations. The influence of Freud's theory of personality is reflected in all strains of structuralism. This is nowhere more evident than in the work of Levi-Strauss, who has stated time and again the essentially unconscious nature of structure, emphasizing as well that one cannot derive an accurate description of a deep structure by simply asking one's informants. Levi-Strauss (1966b) has admitted that there may be from time to time, in all societies, exceptional persons who consciously develop structures that only later become unconsciously embedded, as it were, in the structural fabric of the society. Here he is merely repeating the insight of Paul Radin (1927), who made the distinction between a "thinker" and a "man of action," the latter being the predominant type in society but the former being found at some time or another in all societies.

Biogenetic Structuralism

We have offered this brief summary of the major tenets of structuralism in order both to give the reader unfamiliar with the movement a modicum of background in the subject and to lay a framework from which to contrast orthodox structuralism with our own position, *biogenetic structuralism.*

The major ontological assumption upon which biogenetic structuralism is founded is that *there exists no reality intervening between the central nervous system and the environment.* The corollary is that all other presumed levels of reality have analytic status only. Thus, when philosophers speak of "mind," psychologists speak of "personality," American anthropologists speak of "culture," and sociologists and social anthropologists speak of "society," they are referring to patterns abstracted from behavioral (or introspective) equivalents of internal brain processes. Behavior viewed from our perspective is the synthesis derived from the dialectic between the brain as thesis and environment as antithesis.

This axiom leads to a number of instructive contrasts between orthodox structuralism and biogenetic structuralism. In the first place the heavy reliance by orthodox structuralists upon deduction of structures from observable behavior has led often to the reification of those structures to an ontologically epiphenomenal level of reality. In other words, the analyst starts with social phenomena and by critical examination of the data and formulation of a model that logically accounts for the data, he works back from the observed to the structural. Unfortunately, the locus for these structures all too often turns out to be a subjective but unconscious (and hence unverifiable) "mind" or "reason." The level of structures becomes remarkably similar to νοῦς in the Platonic sense. This results, of course, in a serious defect in any theory that purports to be scientific. For whether we like it or not, science is committed to reductionism. To permit νοῦς or the δεμιουργος, or "a realm of structures," to exist as explanations of observed regularities only serves to perpetuate a dualism more inexplicable than the problem it is constructed to solve. In actual practice, the question of the reality of structures is either avoided outright, or is discussed in a fashion that amounts logically to the same thing. To deduce a structure using a logic that assumes the truth of essentially epiphenomenal postulates,

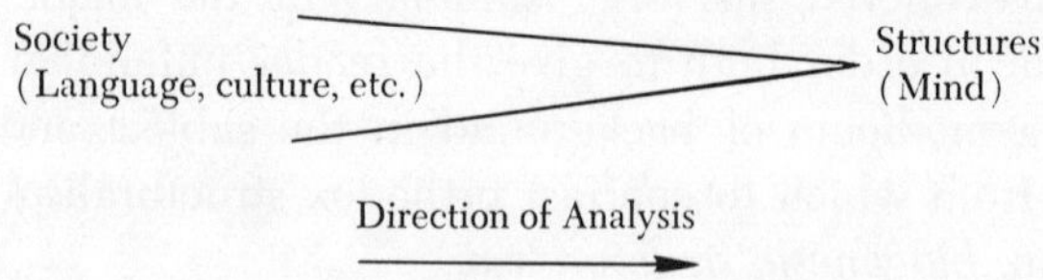

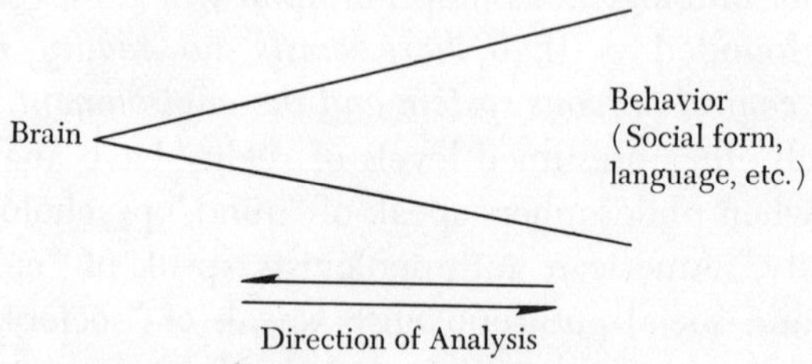

and then to assert that their reality lies in the brain (as Chomsky has done), is not to discover neurophysiological bases for the phenomenon in question. It is logically to beg the question. Biogenetic structuralism explicitly rejects the Neo-Platonic quality of structures

—that is to say, the "superorganic." Put schematically, whereas orthodox structuralism works deductively back from the social to the mental, biogenetic structuralism works by inductive-deductive alternation (see chapter 6) with a heavy emphasis at the present time on induction.

In more practical terms, we propose to analyze real neuroanatomical-neurophysiological relationships, i.e., the neural organization of the brain, as the basis for the "mental" structures considered to be universal. Katz (in press) has noted: "What is remarkably interesting about social and cultural universals is that until recently they were based on a highly simplistic conceptualization of fundamental human biology. In general, universal patterns have an admitted biological basis, but the level of sophistication of knowledge of the biological factors has precluded many important generalizations." At first what we are undertaking may seem absurd or even impossible both to social scientists and to neurophysiologists. Social scientists may feel that if so much vital information relative to higher forms of human behavior has been available it would have become current long before this. A brief consideration of the compartmentalization of knowledge through institutionalization of disciplines in the academy, especially the artificial wall erected between the "hard" and "soft" sciences, ought to persuade one that this need not be the case. The neurophysiologist on the other hand may feel that even his most spectacular results in terms of cerebral control of behavior are stereotyped and impoverished compared with the rich and varied panoply of data on human experience evident in the behavioral sciences, especially the extreme diversity of behavior throughout the world that has been revealed by anthropologists. What the neurophysiologist is *not* aware of is that this dazzling human diversity is only the by-product of the anthropologist's attention to superficiality. Scratch the surface and one finds that this diversity represents "transformations" of finite and simpler, and more "stereotyped" cognitive elements, elements which will be explored in detail in the final chapters of this book. Thus the apparent schism between results in the neurophysiologist's laboratory and descriptions of "higher human social functioning" may not be as vast as it seems at first blush.

In the second place, orthodox structuralism is, as usually formulated, non-evolutionary. It includes (i) no attempt to discuss the evolution of the capacity for structuring, (ii) no attempt to relate crucial events in the evolution of the human species to structural phenomena, and (iii) no attempt to explain change in structures from simple forms to more elaborated forms.

Biogenetic structuralism, on the other hand, represents a merger of evolutionary theory with structuralism. It permits examination of structural-functional relationships in the central nervous system as they correlate with structuring capabilities evident in behavior. Finally, it is a position that is vitally interested in the micro-causal events resulting in change in structures. It is in the attempt to found structuralism on a meaningful physical base and to answer questions concerning the evolution of cognitive structures through the course of hominid development that the present short work is written. We certainly do not claim to know all the answers; we merely hope to suggest how to ask the significant questions and, perhaps, demonstrate a few appropriate answers as well.[4]

We are not so blind nor so naive, as to believe that all the problems in the study of sociocultural phenomena can be resolved tomorrow by invoking a new god—the neurological sciences. We are aware that ignorance in these disciplines is great, especially as regards the higher cortical functions. Nevertheless, our ignorance in these disciplines is far from total. The neurological sciences have been constantly accumulating information about higher cortical functions since the beginning of the latter half of the last century. With a few notable exceptions (see Count, 1973), virtually none of this information has been incorporated into the body of anthropological, sociological, and linguistic theory. It is our intent to open the door to the possibilities of cross-fertilization by a comprehensive theory derived from insights of the neurological sciences, cognitive psychology, linguistics, human paleontology, and social-cultural anthropology.

The strength of biogenetic structuralist theory, as we have sug-

[4] In addition to the applications of biogenetic structural theory contained in this book, we plan to publish more complete analyses in subsequent works. In particular, we intend to examine the nature of pan-specific ritual behavior and its manifestation in man.

gested, lies in its capacity to explain much of the cognitive and structural aspects of classical structuralism by lodging structures squarely in specific cerebral structures and functions. It furthermore combines this strength with an evolutionary perspective that allows one to consider the evolution of such cerebral structures in light of the sweep of hominid evolution and the probable adaptive consequences of each major structure. Let us emphasize that, although biogenetic structuralism may not present all the answers one would wish, its limitations are limitations of basic knowledge and not the limitations imposed by logically fallacious assertions. It is amenable to the deduction and testing of hypotheses in a way that classical structuralism often is not.

The term "structure" will be used in three distinct senses in this book. First, structure may refer to the phenomenal arrangement of parts in a social, cognitive, or linguistic system. The words "superficial" or "surface" will usually appear to denote this usage. Second, structure may refer to those unconscious cognitive elements and relations between elements that classical structuralists conceive as underlying behavior. This is deep structure in the Levi-Strauss–Chomskian sense. Third, structure will be used to refer in a very literal way to neuroanatomical configurations in the brain. Such neurostructures may be gross as when we consider fiber tracts connecting various parts of the brain, or they may be ultramicroscopic as when we speak of presumed molecular changes facilitating short-term memory. Usually we will be speaking of probable dendritic-axonic-synaptic configurations when we use the term in this sense. At every point possible we will attempt to relate structure at all three levels.

At the outset, we wish to state that we hold the utmost respect for the endeavors of the orthodox structuralists, and especially for the work of Levi-Strauss. When we find it necessary to criticize certain aspects of structuralism, we are motivated simply by our firm conviction that unless the system is recast into a form conducive to the development of an explanatory theory, the hope for a scientific basis for the social sciences is bleak indeed.

Furthermore, we wish to exclude the work of Jean Piaget from the criticisms of structuralism presented above, and in subsequent chapters. The tenets of Piaget's theories, we feel, are more "biogenetic" than "orthodox." Of major importance is his concern for the biological foundations of knowledge and the ontogeny of structures. *Ontogeny* means the path of development of structures in the individual member of society, and much of Piaget's voluminous work has been directed to unraveling the nature of social and psychological variables that facilitate and inhibit this process (see Piaget, 1971, 1973). We will not lay heavy emphasis on the process of ontogeny in this book. Our avoidance of the subject is not meant to reflect its importance to biogenetic structuralism. In fact, it is centrally important and the crux of our discussion of the empirical modification cycle (chapter 4). This omission is in the interests of simplifying the discussion so that a coherent theory may emerge. Readers who are interested in the ontogeny of structure are refered to the works of Piaget and to Harvey, Hunt, and Schroder (1961).

Finally, readers who are expert in any one of the disciplines upon which biogenetic structuralism is based may feel that their particular area is not treated in sufficient depth. Our goal, however, is to present a coherent theory that will be of interest to readers who span the spectrum from the neurological sciences to the behavioral and social sciences. The presentation of essential background material in each discipline will be made without the presumption that the reader has any background in that field beyond that of the average educated person. In the following chapters each area will be treated only in detail sufficient to move the development of the theory forward. This brief book is not meant as the last word. We hope it will be only the first.

References

Bergson, Henri. 1907. *L'Evolution Creatrice.* Paris: Felix Alcan

von Bertalanffy, Ludwig. 1968. *General System Theory. New York*: George Braziller.

von Bertalanffy, L., and A. Rappaport (eds.). 1956—71. *General Systems. Yearbook of the Society for General Systems Research* (16 vols.), Washington, D.C.

Count, E. W. 1973. *Being and Becoming Human.* New York: Van Nostrand.

Durkheim, Emile. 1972. *Selected Writings* (ed. by A. Giddens). Cambridge: Cambridge University Press.

Goodenough, W. H. 1956. "Componential Analysis and the Study of Meaning." *Language,* 32:195-216.

Harvey, O. J., D. E. Hunt, and H. M. Schroder. 1961. *Conceptual Systems and Personality Organization.* New York: John Wiley.

Hertz, Robert. 1909. *Death and the Right Hand* (the 1960 edition). Aberdeen: Cohen and West.

Katz, S. H. 1972. "Unity and Diversity of Man from the Point of View of Social and Cultural Anthropology." Presented at Seminar on Unity and Diversity of Man, Fondation Royaumont, Paris, September. Also in *The Unity of Man,* Paris (in press).

Levi-Strauss, Claude. 1966a. *The Savage Mind.* Chicago: University of Chicago Press.

——1966b. "The Future of Kinship Studies: The Huxley Memorial Lecture." *Proceedings of the Royal Anthropological Institute,* 1965: 13-22.

——1967. *Structural Anthropology.* Garden City: Doubleday.

——1969a. *The Elementary Structures of Kinship.* Boston: Beacon Press.

——1969b. *The Raw and the Cooked.* New York: Harper Torchbooks.

Lounsbury, F. G. 1964. "A Formal Account of the Crow—and Omaha—Type Kinship Terminologies." In *Explorations in Cultural Anthropology, Essays in Honor of George Peter Murdock* (W. H. Goodenough, ed.). New York: McGraw-Hill.

Mauss, Marcel. 1925. *The Gift* (the 1967 translation by I. Cunnison). New York: Norton.

Piaget, Jean. 1971. *Biology and Knowledge.* Chicago: University of Chicago Press.

——1973. *The Child and Reality.* New York: Grossman.

Radin, Paul. 1927. *Primitive Man as Philosopher* (the 1957 edition). New York: Dover Publications.

de Saussure, F. 1959. *Course in General Linguistics* (ed. by C. Bally and A. Sechehaye). New York: Philosophical Library.

Stent, G. S. 1972. "Prematurity and Uniqueness in Scientific Discovery." *Scientific American,* 227(6):84.

Whitehead, A. N. 1929. *Process and Reality.* New York: Macmillan.

II

Cerebral Organization and Kladogenesis

In the human line of descent it is quite reasonable to regard the increase of the brain, the development of a motor area of speech, the prolongation of the juvenile phase, the erect posture in walking, and the tool-using hand as essential selective advantages. Human descent, then, was largely necessitated, and is no exception to the general processes of evolution and evolutionary progress which have affected other animal lines. In human phylogeny the typical characteristics of anagenesis are exemplified perfectly well: increased complexity and rationalization, improved plasticity of structures and functions, independence of surroundings, and autonomy. After all, it was a brain victory that brought man into the ruling position of the world, as it was a brain victory when the creodonts were defeated by the carnivores and the prosimians by the monkeys and apes.—BERNHARD RENSCH, *Evolution above the Species Level*

THERE HAS BEEN considerable controversy in recent years over whether a more complex behavioral adaptation of an evolving species is related in a simple fashion to the development of a bigger brain or whether it follows the development of more complex neural organization independent of brain size. This has become a very lively issue especially in respect to the development of human cognition and social behavior. Simply put, the question is: Did human higher cognitive and social functions evolve because of the tremendous increase in cranial capacity during the Pleistocene? Or, contrariwise, did these functions, at least in rudimentary form, evolve while the size of the brain in relation to the size of the body was equivalent to that of a present-day ape? If the latter was the case, then those faculties that we call higher cognition owe their development to the evolution of complex neural reorganization. If such be the case, the subsequent increase in brain size can only be seen as an elaboration and development of previously evolved neural structures.

Both sides of the question have some merit, but, stated in this way, the problem appears somewhat simplistic. On one hand, for example, neural organization must underlie such faculties as "intelligence" or problem solving ability *within* any given species. As we shall demonstrate below, there is enormous variation in the brain size of present-day man without any measurable intelligence (or other) differential as a function of size. On the other hand, when one compares species or, more impressively, when one compares genera there seems to be a strong (although not absolute) relationship between *average* brain size and species competence (or, if you will, "intelligence"). If we assume that both these statements are true we must inquire into the possible relationship between neural organization and brain size. Before we embark on a consideration of these thorny problems, it might be best for us to preview their conclusions so that readers unfamiliar with the often tangled arguments can retain a point of reference throughout the chapter.

We support the position that it is probably neural reorganization that underlies any behavioral (or cognitive) change in an evolving species. Since such changes are often crucial in adaptation to altered environmental circumstances, it can be seen that neural reorganization is basic to the evolutionary process. Furthermore, neural reorganization can take place without any increase in brain size whatsoever. Conversely some increase in brain size can take place without substantial increase in neural complexity. This having been said, however, we also contend that the evolution of bigger brains (according to the kladogenetic model described below) tends to increase the *probability* of complex neural reorganization. Thus increasing brain size among primates over time indicates a high probability of an overall increase in "intelligence." We would define "intelligence" in the evolutionary sense to mean the capacity of a species in relation to any other species to differentiate and integrate perceptual information into adaptive neural models of their environment (see chapter 4; also see Jerison, 1973, p. 4, for a similar definition of intelligence).

With regard to the evolution of higher cortical functions in the hominid line, we shall present evidence in chapter 3 that these functions (abstractive ability, complex problem solving, possibly even language) probably developed as a consequence of neural reorganization at the *prehominid* level, and thus were not a result of the marked increase in brain size during the Pleistocene. This latter increment undoubtedly resulted only as a progressive elaboration of the systems laid down at the prehominid level. If the neural reorganization underlying higher cortical functions is related to an increment in brain size, it is undoubtedly related to the increment in the development of the brain from the simian (or monkey) level to the pongid (or ape) level. If this is so, the question that now arises is what is the significance of the vast increase in cranial capacity between *Australopithecus* and *Homo sapiens?*

If all the basic neural systems underlying higher cognitive functioning are already present in *Australopithecus,* which had a cranial capacity approximately equal to that of modern apes, what does the marked increase in cranial capacity during the course of hominid evolution mean? The answer appears to lie in making a basic distinction between the capability to perform higher cortical functions in a *rudimentary* way based on newly evolved neural systems and the spontaneous and highly efficient functioning of those systems based on their evolving neural elaboration. Although it is absurd (as we shall see in chapter 3) to deny some sort of primitive culture (or even language) to *Australopithecus* just because he had a small brain, it is just as absurd to say that *Australopithecus* was as intelligent as *Homo sapiens* simply because he may have possessed the basic neural structures that underlie higher cognitive and social functioning. It seems highly probable that man's functional intelligence increased as these basic neural systems became progressively elaborated.

Although it is only an inference, we maintain that it is reasonable to suppose that increased cranial capacity during hominid evolution generally represents increased spontaneous and functional intelligence insofar as it represents significant neural elaboration of structures developed at the prehominid level. Detailed arguments in support of this position will be presented in chapter 3. Suffice it to say at this point that, although neural reorganization is absolutely fundamental to the evolutionary process, this by no means negates the importance of the study of cranial capacities of fossil specimens. There are at least two reasons why such studies are of tremendous importance in the reconstruction of man's evolution. In the first place, in the context of primate evolution as a whole, the evolution of bigger brains was probably an important determinant in neural reorganization that in turn underlies behavioral changes upon which natural selection often acts. Second, with respect to hominid evolution in particular, the evolution of bigger brains probably indicates not so much the development of new cerebral capacities as the progressive, efficient, and spontaneous function of those capacities already present at the australopithecine level. In this sense increased cranial capacity in the hominid line may very well indicate increased functional intelligence.

Having previewed our overall position let us now consider in detail some of the problems involved in analyzing the fossil evidence of hominid evolution, and particularly the problems in determining and interpreting cranial capacities in fossil specimens.

As already noted, it is widely held that the present complexity of human cognition and social behavior is in some way linked to a remarkable increase in the size and cognitive capacity of the brain (see Washburn and Howell, 1960; Tobias, 1971, pp. 114 ff.). Tobias (1971) has written that evidence of increased brain size in the hominid line "is the most strikingly sustained trend shown in the fossil record and, hence, in the morphological evolution of the *Hominidae*. . . . The trend toward increased brain size is the most continuous, long-lasting, and marked hallmark of hominization" (pp. 114-115).

If the finds by Leakey (1973) are dated correctly this expansion began at least 2.9 million years ago. Leakey's cranium specimen (KNM-ER 1470) has been provisionally assigned a cranial capacity of over 800 cubic centimeters, well above the australopithecine range (roughly 450 to 650 cubic centimeters). This supports the attribution of the genus *Homo* to these specimens, thus pushing back the appearance of the genus two million years. Prior to this discovery the first notable increase in cranial capacity was thought to have occurred with the first appearance of *Homo erectus*, possibly one million years ago at the earliest. KNM-ER 1470 may represent a transitional form between the australopithecine level and *Homo erectus*. However, if KNM-ER 1470 is indeed 2.9 million years old, *Homo* could not possibly have descended from the later australopithecines who were around well after that date. Undoubtedly both *Homo* and the later australopithecines evolved from an early australopithecine or australopithecine-like form. To suppose otherwise would demand drastic and highly improbable evolutionary jumps within the hominid line.

We wish to emphasize, at this point, that we do not want to embroil ourselves in futile arguments over the precise line of descent of *Homo sapiens*. Whether, for example, most of the *Homo erectus* skulls represent man's direct ancestor or a collateral branch is not germane to the considerations in this book. The overall increase in cranial capacity and the differential development of brain parts within the hominidae are our concern. Thus, without specifying particular genera and species, we will speak of the following sequence of "levels" when discussing brain evolution throughout this book: (1) protohominid level; (2) primitive hominid level (e.g.,? *Ramapithecus*); (3) early australopithecine level; (4) early *Homo* level (e.g.,? KNM-ER 1470); (5) *Homo erectus* level; (6) early *Homo sapiens* level; (7) modern *Homo sapiens*.

Let us now return to our discussion of increasing cranial capacity within the hominid line. This trend culminates in essentially modern dimensions, at the latest around 300,000 to 180,000 years ago with the Swanscombe and Steinheim skulls and at the earliest with Sinanthropus 10 from Choukoutien.[1] Few would doubt the fact

[1] Our own calculations based upon published data. Dating of the initial terminus of any such sequence should be considered extremely tentative.

that such an increase did occur in the evolution of the hominids. However, the nature and the condition of the fossil data raise a number of problematic questions that have yet to be satisfactorily answered: (i) What is the relationship between increased cranial capacity and increased cognitive capacity? (ii) What evolutionary mechanisms caused the increase in hominid cranial capacity? (iii) What can we learn about the evolution of the brain from a study of endocranial casts? (iv) What were the correlative causal effects of increased cranial capacity and brain size in terms of social interaction, adaptive patterns, and conceptual modeling? The present chapter will pose at least partial answers to the first three questions. The final question must wait for a later chapter.

Cranial Capacity and Increased Intelligence

There are a number of problems in utilizing the data on cranial capacities extracted from the fossil record. These problems are often glossed over. One problem is in determining the relationship between cranial capacity and intelligence. It is assumed by many that selection at some point began to favor an increase in hominid intelligence, and that this selection resulted in ever-increasing brain size. Brain size, common sense will tell us, has something to do with the capacity of the brain case. The rub comes in that there seems to be no simple and clear-cut correlation between intelligence and brain size (see Buettner-Janusch, 1966, p. 146; Holloway, 1968a, p. 170). To make matters worse no simple relationship exists between brain size and size of brain case although this is less a problem in mammals than for lower vertebrates (Jerison, 1973, pp. 26, 84). Even among modern *Homo sapiens*, the range of cranial capacity will vary 400 centimeters on either side of the population mean with no apparent effect upon level of "intelligence" (Harrison, 1964, p. 210). It might also be noted from the summary mean and range of cranial capacities for the australopithecines, *Homo erectus*, and *Homo sapiens* in Figure 2.1 that the *mean* capacity, not to mention much of the range, of *Homo erectus* is within the functional bounds

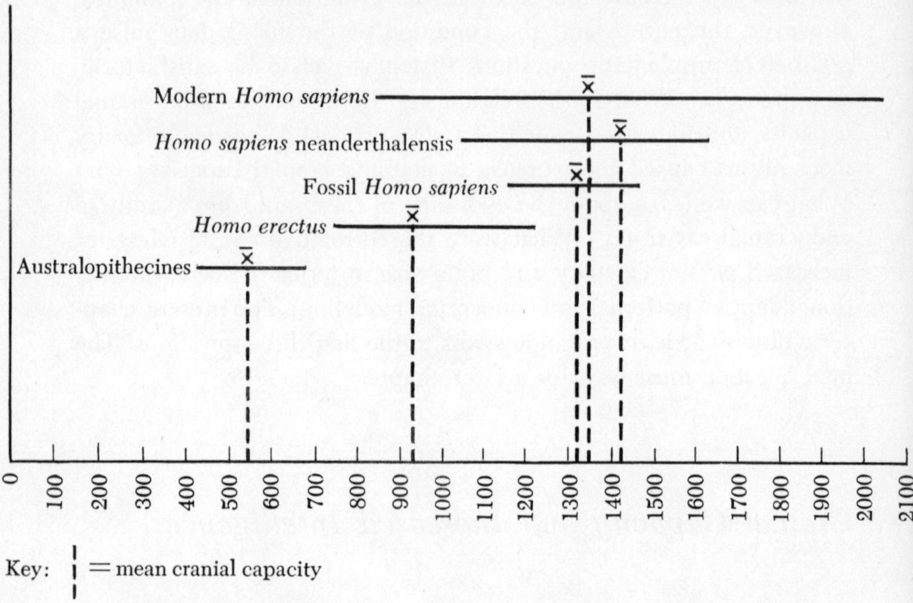

Figure 2.1. Comparative Range and Mean Cranial Capacities, The Hominid Line.

of *Homo sapiens* (see Holloway, 1966, 1968a, 1968b).[2] Does this mean that erectus was as intelligent as modern man? Many authorities would wish to deny this conclusion.

Another methodological problem plaguing researchers interested in the evolution of the human brain is the question of precise measurement of cranial capacity. Authorities are presented with everything from a partial parietal to a complete cranium, and yet they are asked to make estimates of cranial capacity that are compa-

[2] The data from which these and other computations are based were extracted from Tobias (1971), Day (1965), and Coon (1962). The categories presented in Figure 2.1 are comprised of the following specimens: (i) australopithecines, subsuming *A. africanus* (Taung; Sts 5, 60, 19/58, 71; MLD 37/38), *A. robustus* (SK 1585), *A. boisei* (OH 5), and *Homo habilis* (OH 7, 13, 16); (ii) *Homo erectus* (*H. erectus* I, II, III, IV, VI, VII; *H. erectus pekinensis* II, III, X, XI, XII; Lantian specimen; OH 9); (iii) Fossil *H. sapiens* (Swanscombe, Steinheim, Fontéchevade); (iv) *H. sapiens neanderthalensis* (Solo 1, 5, 6, 9, 10, 11, Saccopastore 1, 2, Krapina C, D, Ehringsdorf, La Chapelle, La Ferrassie I, Neanderthal, Spy 1, 2, Circeo 1, Le Moustier, Gibraltar, La Quina, Gánouce, Tabun 1, Skuhl 4, 5, Saldanha, Broken Hill).

rable across the range of fossil specimens. It is obvious that estimates of cranial capacity for complete skulls are the most accurate. In general, the most ancient finds tend to be the most incomplete, thus introducing a potential systematic bias into statistics derived from the record. How then is the capacity to be estimated? The brain case of a fairly complete skull is usually filled with mustard seed or shot and the quantity of the medium thereafter measured. The measurements are obtained in a variety of ways: by the volume of the seed or shot itself, by the liquid displacement of the medium, or by utilizing a formula (usually Tildesley's formula; see Montagu, 1951) for conversion of the weight of the measuring material to volume (Harrison, 1964).

However, many specimens making up the hominid fossil record offer incomplete or nonexistent evidence of cranial capacity. When a portion of the brain case is available, the cranial capacity of the once complete skull is estimated in a variety of ways. The more explicit are based upon one of a number of formulas, the most common of which is the Pearson formula (Montagu, 1951). Other measures have been derived from at least four other formulas (see (Jones, 1929). The use of formulas is by no means the only way cranial capacities have been estimated from partial skull fragments. Tobias (1963) uses a method by which he makes silhouettes from the parietals and, hinging them together, determines a range within which the actual cranial capacity must have lain. Some estimates have been based upon measurements of an artist's reconstruction of a total skull from bone fragments.

The difficulties inherent in estimating cranial capacities are manifest in some of the work of Holloway (1972). His careful determinations of certain australopithecine endocasts yield occasionally fairly significant differences from the estimates of previous and likewise careful workers. This is particularly true of his calculation of the cranial capacity of the Taung baby (404 cubic centimeters, calculated adult size 440 cubic centimeters) as compared with previous calculations (525 to 600 cubic centimeters). In reference to this discrepancy Holloway (1972) notes,

> The rather large difference between the previous volume for the Taung child and my own reconstruction is a matter of correctly finding the most reasonable midsagittal plane, and adding a frontal portion

> which matches the matrix still imbedded in the facial fragment. Of course, one must assume perfect symmetry between hemispheres when arriving at the total volume. (p. 192)

With reference to another australopithecine endocast Holloway (1972) states:

> The STS 71 cranium is highly distorted, and the endocast portions taken from it were likewise distorted. The reconstruction attempted to rid the endocast of that distortion, and so provide a final form that bore closer resemblance to an australopithecine brain. Obviously, this was a matter of some "feel," and I hope that my familiarity with these specimens has led me to make the proper decisions. (p. 193)

Not to belabor the point, the obvious methodological question is if all other parameters are assumed to be equal, just how accurately comparable are the estimates derived from disparate techniques? When we make a comparison based upon cranial capacities determined by liquid displacement of a naturally formed brain case cast on the one hand and those determined variously by external measurements of an artist's reconstruction or by Tobias' silhouette method on the other hand, how confident are we expected to feel? The answer is a depressing uncertainty. We are, therefore, compelled to emphasize three difficulties relative to the measurement of cranial capacity: (i) One need only look closely at the literature to find cases of conflicting cranial capacity estimates, often accompanied by a theoretical proposition hinging upon the critical measurement. (ii) There is doubt as to how accurate the various formulas are in estimating the cranial capacity. All the formulas mentioned above require *external* measurements and all, save one, use length, height, and breadth. In his remarkable study of craniometry, W. W. Howells (1957) found these three factors to be the most important in describing and explaining external cranial variation among humans. However, he found that only one of these, the breadth factor, correlated very highly with cranial capacity. For this reason, he has suggested that cranial capacity be measured directly as it has a "remarkable variation independent" of any of the ten factors he examined. (iii) And the difficulty with loose estimates like Tobias' silhouette technique and measurements, however accurate, from artists' conceptions is that they depend too heavily upon an appeal to authority and are too amenable to distortion due to subconsciously biased presuppositions.

A final methodological problem pertaining to the evolutionary study of hominid brain stems from a characteristic of the fossil record itself. Once we go back much beyond the Upper Pleistocene, fossil hominid specimens yielding data on cranial capacity become increasingly scarce and spatiotemporally clustered. The Middle Pleistocene finds at Choukoutien and in Java all seem to cluster in several definite geographical and temporal nexuses. The specimens from South and East Africa represent an even longer period of evolutionary time, and yet are spatiotemporally "clumped." In short, we have nothing approaching a random sample of cranial vaults at any point in time until possibly the Upper Pleistocene. Thus, when we state with no qualifications that the mean cranial capacity of *Australopithecus* was something less than 550 cubic centimeters, we are making a fairly sweeping assumption about the sample of fossil finds upon which that estimate is based, namely, that the sample of specimens we are able to measure are truly representative of the australopithecine population throughout the duration of its existence. The best we can say about the evidence at hand is that any particular spatiotemporal cluster of specimens probably groups around the mean cranial capacity for that local population. Yet even this assumption remains questionable. Also, it must be remembered that among modern man the mean cranial capacity for various local populations will differ greatly (Montagu, 1951; Harrison, 1964).

As depicted in Figure 2.1, the mean cranial capacity for all measurable specimens of *Homo sapiens neanderthalensis* is approximately 70 cubic centimeters *larger* than the mean for modern *Homo sapiens*. This computation has been widely disseminated and has been interpreted as evidence of larger brain size, and even greater intelligence, of classical Neanderthal Man. (It will also be noted in Figure 2.1 that the mean for early fossil *Homo sapiens* is some 30 cubic centimeters smaller than that of modern man.) Conclusions that assert characteristics of the Neanderthal population on the basis of measures of central tendency derived from the samples available commit the most fundamental of statistical errors. It seems more likely that the means for the fossil populations represent discrepancies due to sampling error rather than actual differences in population characteristics. As a matter of fact, the way the ranges of the Neanderthal and early *Homo sapiens* samples position with reference to the mean for modern man is much as we would expect *if* the true mean for the hominid line had remained substantially the same for at least the last 200,000 years. This would be the case *assuming* that means for samples as large as these two fossil clusters would tend to group fairly closely around (but not to be coterminous with) the true mean.

In chapter 3 we intend to analyze the adaptive functioning of the human neocortex into a number of interrelated subsystems, each of which can be seen to have adaptive capabilities considered alone, but the combination of which is seen to provide the neural basis for language and social behavior. Such an undertaking cannot properly avoid a consideration of these systems within the context of hominid evolution. This obviously requires using the very fossil record we have been critically evaluating above. Yet, as we have seen, this presents us with difficulties. On the one hand we have the school of "paleoneurology" exemplified by Krantz (1968, 1970) that attempts to correlate early hominid behavior with gross brain size (indicated, of course, by cranial capacity). This school, which until recently completely dominated human paleontology, has often denied the possibility of culture to the australopithecines, using small brain size upon which to found their judgment. On the other hand, after pointing to some of the difficulties with the study of brain size described above, Holloway (1968b, pp. 123 ff.) proposes neural reorganization, rather than development of brain size per se, as the primary factor in hominization.

Holloway states that neural reorganization ". . . means that quantitative shifts between components or substructures of the brain, as measured in terms of area or volume, have taken place under natural selection such that the outputs of the systems are different between the species. By shifting interactions between components in quantitative ways, the product of the whole is altered" (1968a, p. 175). In his fairly complete treatment of evidence for neural reorganization in the process of hominid evolution, Holloway suggests that such reorganization has occurred not only at the gross anatomical level, both in the neocortex and in subcortical regions, but also at the microstructural level with changes in proportion of cell types, differential cell densities, and the degree of dendritic branching (see Holloway, 1968a). Probably most important to our present discussion, he suggests that, even if truly accurate methods existed for the comparison of cranial capacities and brain sizes between species, the measures would still not be strictly comparable as, metaphorically speaking, one would be comparing apples and oranges: "A comparison of cranial capacities of extant forms, or those based on endocasts, will not be a comparison of equal units. One cubic centimeter of chimpanzee cortex is not equivalent to one cubic centimeter of human cortex, nor is it likely that any equivalent measure can be found" (Holloway, 1968a, p. 177).

Our own position represents a compromise between these extreme positions (Krantz and Holloway). We agree that neural reorganization did result in the behavioral changes responsible for the manlike social behavior of the early hominids, and that this occurred *without an initial increase in brain size* over that of their prehominid ancestors. However, it seems obvious to us that some neural reorganization accompanies any shift in adaptive behavior, regardless of the species involved. As Jerison (1973) has said:

> In recent years there has been some emphasis on the reorganization of the brain in "higher forms" or in man. That emphasis is misplaced. It is likely that as we learn more about the wiring diagrams of various brains we will recognize more and more differences. These differences should impress us no more than the behavioral differences that are much more easily observed. . . . We should expect adaptations to various niches to be made possible by the evolution of appropriate structures and functions, and reorganizations of the brain would be no more than the neural equivalents of species-specific behavior patterns. (p. 81)

We would argue, furthermore, that progressive and differential elaboration of various cortical systems in the hominid brain resulted in increased efficiency of certain systems. This increased efficiency was, to a certain extent, reflected in elaboration of various parts of the brain and hence discernible by studying endocranial casts. In other words, consideration of the development of various parts of the brain in relation to adjacent structures as represented in endocranial casts, may give us a clue to the differential rates of evolution of various cortical systems, and hence of their relative efficiency in terms of the behavior of a given hominid species or subspecies. We would agree with George Sacher (1970, p. 246) that much useful information about the evolution of the central nervous system of man and the other primates may be obtained by a close analysis of allometric relations between subsystems of the brain. Such data have in fact been offered by Stephan, Bauchot, and Andy (Noback and Montagna, 1970), as well as Jerison (1973).

We have shown at the outset that the relationship of brain system and subsystem function, on the one hand, to the fossil indication of increased skull size, on the other, is very complicated indeed. There are scholars who contend that changes in hominid cranial morphology over the last two million years is primarily due to heavy selection in favor of increased intelligence. We wish to suggest that the picture is much more intricate than previously depicted, involving selection for both body size and intelligence in an extremely complicated pattern.

Kladogenesis

Of the more sophisticated and compelling explanations of the increase in hominid brain size and complexity is that offered by the German school of evolutionary biology, particularly by Bernhard Rensch (1959). Several of the pivotal concepts and relations utilized in Rensch's work can be applied to yield a more complete understanding of the evolution of human cognition and social behavior. We will not attempt here either to expound on Rensch's theory at length or to defend his formations. And we are perfectly aware of the largely *post hoc* nature of this exercise. Nonetheless, the formulation has much to commend it; certainly it is superior to most of the historical speculations that have gone before.

Rensch (1959, pp. 210 ff.) presents an impressive array of data showing a clear tendency for most higher phyla to develop forms with increased body size. This tendency is one of many that characterize the process of phyletic evolution Rensch terms *kladogenesis.* Kladogenesis may be defined as a set of predictable regularities in the evolutionary change of forms when analyzed above the species level. These regularities occur when phyla branch out or expand in reaction to relatively rapid change in the environment (p. 97). Change does not happen at the same rate through time, but rather in fits and starts. At the beginning there is a period of radiation, or an "explosive phase," in which a relatively large number of subphyletic forms (e.g., genera, species) emerge to fill new ecological niches. After the period of radiation, there is a period of specialization during which fewer new forms emerge, many of the initial forms become extinct, and some of the initial variants survive and specialize. It is during the period of kladogenetic change (or nonrandom elaboration of new forms) that the phyla generate their larger (as well as their smaller) subphyletic members.

A significant number of nonrandom tendencies have been isolated in the study of kladogenetic change. Important to the present study are the relational regularities between increase in total body size on the one hand and increase in constituent organs on the other. Particular organs are found to change in size at rates proportionately greater than, less than, or the same as the change in size of the entire body. This may be termed respectively positive, negative, or neutral *allometry.* For example, the change in the length of the intestines is much greater than the overall change in body size for the reason that while body size increases by the cube, the capacity of the intestines to absorb nutrients increases only by the square (efficiency of the intestines being a function of its internal surface area). The organs must more than double in length to accommodate doubled body size. The intestines thus exhibit positive allometry in relation to body size.

To return to the main point, the average absolute size and weight of the brain in a form increase as an allometric by-product of selection in favor of larger body size (Rensch, 1959, pp. 302 ff.). It is found that the brain in all higher forms for which there is evidence has a negative allometric relationship to body size (see Holloway, 1966, p. 113, for evidence of this among the hominids). Despite this fact, the brain does increase in size, and it does so not only by a multiplication of the number of constituent cells (including neurons and other "supportive" cells) but also (unlike other body tissue) by an increase in the overall size and complexity of the constituent neurons. As neurons increase in size, their dendritic-synaptic fields tend to become more ramified (Rensch, 1959; see also Jerison, 1973, p. 68). The development of new and specific neural connections becomes ever more probable. Ultimately the enlarged nerve cells may become not only capable of handling greater quantities of information but also of dealing with more different types of association than was previously possible (*á la* Cope's Rule.)[3] Increased size of the brain in a new species may thus result in a qualitative change in the capacity and functioning of the brain (Rensch, 1959, pp. 154, 215 ff.).

We are not asserting, however, a simplistic correlation between brain size and intelligence among members of any particular species. As stated above, kladogenesis is a tendency operating above the species level and probably above the genus level as well. Thus, within a given species (or even genus), it is highly improbable that individuals with larger brains represent more complex neural organization. The argument that increased average brain size is somehow correlated with more complex neural organization becomes persuasive only when one compares genera. Thus we are concerned with quantum changes over long periods of evolutionary time. For example, the range in the cranial capacity of modern man, as we have already stated, is 650 to 2050 cubic centimeters (Figure 2.1), greater than the difference between the mean cranial capacity of modern man and that of *Australopithecus*.[4] What is critical, of course, is the increment in the mean size of the brain over time because allometric increase in brain size is accompanied by an increment in the complexity of the constituent cells. The result is one of many orthogenetic tendencies in the emergence of forms with more complex brains Holloway (1966, 1968a, 1968b) was correct to emphasize the importance of reorganization of interneural connec-

[3] Cope's rule states that a quantitative morphological change in an organ will result finally in a qualitative change in the function(s) of that organ.

[4] Also see Holloway (1968a) for similar conclusions.

tion in hominid neural evolution. As evidence of brain functioning accumulates, it becomes more certain that it is to the complexity of dendritic-axonic configurations, size and exchange capacity of synapses, and interlocking neural tracts that we must look for an understanding of the associative processes. However one cannot ignore progressive increase in cranial capacity, especially when one is considering the whole sweep of primate evolution.

Returning again to Rensch's kladogenetic model, we saw that in the first, or explosive, phase a variety of forms radiated from the original form to fill newly available ecological niches. We also saw a tendency for phyla to develop forms of ever-increasing size with concomitantly larger brains. Yet we know that smaller forms emerge as well. Does it then follow that a *decrease* in body size will lead to a *reduction* in complexity of the brain in keeping with a reduction in size? The answer is that the process outlined above resulting from increased body size may *not* actually be reversed with a decrease in body size. Although we would expect some decrease in absolute size of the brain as a result of decreased body size, we would not expect a decrease in the complexity of the neural organization so long as that organization underlay behavior that is still adaptive. *Selection would act in rigorous opposition to a marked decrease in the associative capacity of the brain* (i.e., the complexity of neural adaptation), *especially in progressively diminutive forms.* Hence what we should find is that smaller progressive forms have roughly the same neural association capability and level of organization as the precedent parent form, whereas larger progressive forms tend to develop more highly advanced associative capacities. From this argument, we would expect both the *africanus* and the *robustus* forms of *Australopithecus* to have roughly equivalent associative capabilities and intelligence.

Unfortunately, the fossil record provides us with no direct evidence of an increase in the size and complexity of hominid neurones. Also, we have no certain evidence of the cranial capacity of *Ramapithecus*, the Miocene primate usually considered on the basis of dentition to be the earliest precursor of *H. sapiens* (Tobias, 1971, p. 114). Furthermore, even if we did have direct evidence of the cranial capacity of this creature, it would shed, as we have said, very little direct light on the nature of its constituent neurons. Yet we feel we are safe in arguing that selection for increased body size did occur in the hominoid line (apes and man), and, hence, we may argue by inductive extension from Rensch's findings that an increase in the size, and more importantly in the *complexity*, of the hominoid brain occurred as a consequence. Certainly this change was occurring by the time represented by the earliest australopithecine finds.

Also apparent from the study of kladogenetic change is a high (though not inevitable) correlation between complexity of central nervous system and complexity of inherent social interaction:

> With regard to instincts we may state that in all small European species of bees and wasps (such as *Andrena, Halictus, Ancristocerus,* and *Symmorphus*) there are no social instincts such as exist in the large species of *Apis, Bombus, Vespa, Polistes,* and so forth. Among ants the small Ponerinae have simpler social instincts than the larger Camponotinae, and the smallest ant species *Monomorium salomonis* . . . show only poor social instincts. Among the Scarabaeidae the large species like *Scarabaeus, Copris,* and *Geotrupes* show complicated instincts in the care of their eggs, and in the construction of breeding pills from dung, subterranean canals, and breeding chambers, whereas the many small species of the related genus *Aphodius* have not developed such instincts. . . . The same holds good for the Silphidae, in which the large species of *Necrophorous* show very complex instincts by burrowing in the carcass and feeding the young larvae, whereas the small species of the related genus *Catops* . . . show no such instincts. (Rensch, 1959, p. 164)

Though Rensch takes most of his examples of this correlation from populations of insects, he did note that the same held in many cases for vertebrates. He found much more decisive and interesting results from the experimental examination of the relationship between increased body size and more effective learning and memory storage:

> Concerning domestic fowl, Altevogt . . . stated that a large race mastered simultaneously more visual discrimination tasks . . . than the dwarf races. . . . The smaller races were quick in learning simple visual discriminations, but the larger types were better in learning more complicated and difficult discriminations, and moreover, were superior in memorizing.
>
> The same results are obtained in the study of mice and rats. In the

> latter species the visual learning process is quicker and the number of tasks finally mastered simultaneously is greater. . . . As to retention and transposition ability, the larger rats were again superior to the smaller mice. . . . Continuing this kind of visual discrimination experiment, we were able to show that a mammal with an absolutely large brain . . . , the Indian elephant, is especially capable of learning and retaining a remarkable number of visual tasks. . . .
>
> Finally, the experiments on the prelinguistic capability of "comprehending" unnamed numbers by Koehler and his school showed that large bird species were better in learning and transposing seen and heard numbers than related smaller types. Generally, ravens were more efficient than jackdaws, and large parrots learned more than small parakeets. . . . (Rensch, 1959, pp. 164-165)

Summary

Thus far we have argued that the hominid line is the result of phylogenetic branching among the Hominoidea. This did not occur only once but many times in the evolutionary history of the superfamily. At each branching there was selection in favor of increased body size. Because of the allometric relationship between the size of the brain and the absolute size of the body, there was a progressive increase in the size of the brain, reaching its most dramatic dimensions in Hominidae. Critical to our understanding of the evolution of hominid cognition is the fact that along with an increase in the absolute size of the brain, there also occurred a concomitant increase in the size and dendritic-axonic complexity of the constituent neurons. This in turn led to greater complexity of dendritic-synaptic configurations and neural tracts.

A point was reached in the kladogenetic process just outlined when a shift—*the critical shift* in our opinion—in the qualitative functioning of the hominid brain occurred. That is, the *quantitative buildup* of neural complexity created by the time of australopithecines resulted in an ultimate *qualitative change* in the associative capabilities of the hominid brain. And it is an examination of this revolutionary neural reorganization along with its cognitive and social ramifications that forms the concern of the rest of this book.

References

Buettner-Janusch, John. 1966. *Origins of Man.* New York: John Wiley.

Coon, C.S. 1962. *Origin of Races.* New York: Alfred A. Knopf.

Day Michael H. 1965. *Guide to Fossil Man.* New York: World Publishing Company.

Harrison, G.A. 1964. *Human Biology.* New York: Oxford University Press.

Holloway, R.L. Jr. 1966. "Cranial Capacity, Neural Reorganization, and Hominid Evolution: A Search for More Suitable Parameters." *American Anthropologist,* 68:103-121.

——1968a. "Cranial Capacity and the Evolution of the Human Brain." In *Culture: Man's Adaptive Dimension* (ed. by M.F.A. Montagu). New York: Oxford University Press.

——1968b. "The Evolution of the Primate Brain: Some Aspects of Quantitative Relations." *Brain Research,* 7(2):121-172.

——1972. "Australopithecine Endocasts, Brain Evolution in the Hominoidea, and a Model of Hominid Evolution." In *The Functional and Evolutionary Biology of Primates* (ed. by R. Tuttle). Chicago: Aldine.

Howells, William W. 1957. "The Cranial Vault: Factors of Size and Shape." *American Journal of Physical Anthropology,* pp. 19-48.

Jacob, T. 1966. "The Sixth Skull Cap of Pithecanthropus Erectus," *American Journal of Physical Anthropology,* 25:243-60.

Jerison, H.J. 1973. *Evolution of the Brain and Intelligence.* New York: Academic Press.

Jones, F.W. 1929. *Measurements and Landmarks in Physical Anthropology.* Honolulu: Bishop Museum.

Krantz, G.S. 1968. "Brain Size and Hunting Ability in Earliest Man." *Current Anthropology,* 9:450-451.

——1970. "On Brain Size and Behavior in Earliest Man." *Current Anthropology,* 11:176.

Leakey, R.E.G. 1973. "Evidence for an Advanced Plio-Pleistocene Hominid from East Rudolf, Kenya." *Nature,* 242:447-450.

Montagu, M.F.A. 1951. *An Introduction to Physical Anthropology.* Springfield, Ill.: Thomas.

Noback, C.R., and W. Montagna. 1970. *The Primate Brain.* New York: Appleton-Century-Crofts.

Rensch, B. 1959. *Evolution above the Species Level.* New York: Columbia University Press.

Sacher, G.A. 1970. "Allometric and Factorial Analysis of Brain Structure in Insectivores and Primates." In *The Primate Brain* (ed, by C.R. Noback and W. Montagna). New York: Appleton-Century-Crofts.

Tobias, P.V. 1963. "Cranial Capacity of Zinjanthropus and Other Australopithecines." *Nature,* 197:743-746.

——1971. *The Brain in Hominid Evolution.* New York: Columbia University Press.

Washburn, S.L., and F. Clark Howell. 1960. "Human Evolution and Culture." In *Evolution After Darwin.* Vol. 11, *The Evolution of Man* (ed. by Sol Tax). Chicago: University of Chicago Press.

III

Cerebral Adaptation and Hominid Evolution

The real differences which exist between the brain of man and that of the higher apes are indeed minimal. There can be no illusion about that. Man is much closer to anthropoid apes in terms of the anatomical characteristics of his brain than these latter are, not only to other mammals, but even to certain monkeys such as *guenons* and macaques.—VULPIAN, *Lecture sur la Physiologie* (1866)

IN THE LAST CHAPTER we proposed a mechanism that explains the evolution of increased brain size as a function of selection for larger body size. We also indicated that it is by no means obvious that increased brain size automatically results in the development of those adaptive functions that we call higher cognition. Indeed, as we shall attempt to elaborate in this chapter, it is not the size, *per se,* that is responsible for the evolution of higher cognition, but rather the complexity and organization within the brain, a position also espoused by Holloway (1966, 1968a, 1968b). The importance of increased brain size is that it greatly increases the *probability* of new neural connections (i.e., neural reorganization). Let us repeat, it is not the brain size, *in itself,* that is responsible for increased cognitive adaptation. As we have also indicated, it is conceivable that once a specific neural reorganization is selected for, and becomes firmly established, the species may evolve toward smaller body size (with concomitant decrease in overall brain size) and still preserve the new neural organization. This assumes, of course, that the behavior resulting from the newly acquired neural reorganization remains adaptive in a changing ecological niche. Obviously the question becomes complex indeed. In the case of man, at least, it is the *complexity* of the neural organization that results in the highly adaptive behavior involving "higher cognition."

The correlation between complexity of neural organization and species behavior is undoubtedly not confined to man. Stephan and Andy's (1966) excellent work on quantitative comparisons of brain structures from insectivores to primates lays the foundation for a more general principle of neural reorganization as the major factor determining behavioral changes in evolving species.

At the risk of some repetition we might briefly present here our position regarding the relationship of brain size and neural organization in the context of hominid evolution.

First of all, we contend that the functions of "conceptualization," perception of abstract causality, and abstract problem solving were at least incipiently present at the prehominid level. The hominid line probably split off because special circumstances selected for these characteristics and firmly established them in a group of pongids, so that these faculties began to operate spontaneously. A similar neural reorganization has been demonstrated to be present in the chimpanzee, but it does not seem to be firmly established and hence does not operate spontaneously. Presumably the increase in brain size from the *simian* (or monkey) to the *pongid* (or ape) level allowed for this neural reorganization. This was not an infallible development, however, since, so far as we know, the reorganization involving the basis for higher cortical functioning took place in only two pongid strains—the protohominid and the protochimpanzee. Only in the former was this reorganization selected for, and thus only in hominids did it become highly developed. The possible reasons why these faculties were selected for within the hominid line shall be considered subsequently in this chapter. Suffice it to say here that in the chimpanzee the essential neural mechanisms were not elaborated, and consequently the cognitive functions alluded to above (conceptualization, abstract problem solving, etc.) never became established as spontaneous behavior. Once the hominid line had split, the brain underwent vast expansion between the australopithecine level and the level of *Homo erectus.* This expansion resulted primarily in increased functional efficiency. However, the potential for the development of new structures and new connections may have resulted in the evolution of the mechanism of speech at this point. We shall return to the evidence for this possible two-stage evolution of higher cortical functions in considerable detail below. The essential point we wish to make here, however, is that the two explosions of brain size (i.e., from simian to pongid level and from pongid to hominid level) greatly increased the probability that new neural connections underlying new adaptive behaviors would develop. Adaptive reorganization of neural structures, however, does not follow necessarily from increased brain size, nor, once established, does it require the maintenance of that brain size.

Before we consider the evolution of cerebral adaptive mechanisms (i.e., the cognitive faculties mentioned above) within the framework of hominid paleontology, we must first consider them as found in *Homo sapiens.* If we wish to understand the evolution of human society in terms of the evolution of human cerebral capabilities, we must first break down "cerebration" into its essential component elements. In other words, we must attempt to analyze the adaptive functioning of the human neocortex into a number of interrelated subsystems, each of which can be seen to have adaptive capabilities considered singly, but the combination of which is seen to provide the neural basis of the social forms of Upper Paleolithic man and beyond. This will lead inevitably to a somewhat artificial analysis of specific neurophysiological systems. We hope to reintegrate the various systems when we consider the genetics of cerebral adaptation in chapter 8.

At this point we will consider only development of the neocortex in man. The neocortex is the outer layer of the brain, the so-called gray matter. Its evolution is the most recent in the phylogenetic development of the brain. It is generally associated with higher nervous activity including the various aspects of cognition. Although Crosby and Humphrey (1941) and Daitz (1953) have demonstrated important hypothalamic[1] and limbic[2] development in man over other primates, it is the development of the neocortex that is of overriding importance in human adaptation to the physical environment through the genesis of what is usually called culture.

The Neurophysiological Mechanisms of Cerebral Adaptation

First of all, we must ask whether we are dealing with a single mechanism or an aggregate of mechanisms. Since the central nervous system functions as a single unit, a single computer, as it were, being constantly reprogrammed and modified by its environment, it is strictly correct to speak of a single plastic adaptive mechanism when discussing higher nervous function. Are we then justified in discussing any single part of the brain (or even two or three interrelated parts) as being the "seat" of any given function (such as abstraction, problem solving, speech)? This question opens up a hornet's nest of controversy that has been raging in the neurological sciences for the last forty years. For the sake of our readers who are not familiar with the issue we will briefly sketch some arguments on both sides and then present the position that we feel is most reasonable. For the sake of simplicity, we will label the extreme groups in the argument as the localizationists and the generalists.

The study of pathological lesions of specific areas of the brain and concomitant loss of certain functions in neurological patients led investigators in the last half of the nineteenth century and the first quarter of the twentieth to postulate that specific functions

[1] The hypothalamus is a small structure at the base of the brain, old in terms of phylogenetic development, that exerts control over visceral activities, water balance, temperature regulation. It is also involved in sexual responses and "fight or flight responses."

[2] The limbic system consists of a number of structures at the core of the brain (e.g., limbic lobe, hippocampus, amygdala) that are known to be intimately associated with the expression of emotion, particularly with the experience of pleasure and pain (roughly areas 34, 27, 28, 23, 24, and adjacent structures on the brain map in Figure 3.1.)

were controlled by specific areas of the brain. Ablation (removal) and stimulation experiments in animals tended to confirm these findings. In other words, any given function was thought to be localized to a circumscribed area of the brain. Eventually the number of "functions" proliferated along with the number of control areas in the brain. One of the most extreme of the localizationists was Kleist (1934), who so compartmentalized the brain into "control centers" as to force one to conceptualize it as an aggregate of separate organs. Kleist's localization chart of the brain is reproduced in Figure 3.1 to give the reader some idea of the extremes to which the tendency to localize functions went![3]

A number of experiments in the late thirties and forties, foremost among which were those of Lashley and Clark (1946, 1947), began seriously to damage the localizationist position. For example, Lashley removed various portions of animal brains after they had been trained to perform certain tasks. He found that decreased competence in performing the tasks was not a function of *where* the lesion was made, but only of how much tissue was removed. Such experiments, accompanied by an intense interest in field theory that was cresting in physics at that time, produced an extreme reaction to the excesses of the localizationist position. From a position that stated that every function had a specific control center developed a counterposition that no brain centers existed at all for higher nervous functions or, rather, that the whole brain governed all functions.

Gestalt psychologists also attacked the localizing position. They asked, for example, how a tune could be recognized as the same even if played in two different keys since the "absolute value" of each corresponding note in the two keys should be totally different if perception were a purely passive phenomenon "registered" in one part of the brain. Obviously something active had to be going on in the "simple perception" of the tune involving many parts if not the whole brain. Furthermore, experiments demonstrated that many

[3] This map serves another purpose in this place. Since the various areas of the brain are numbered according to a fairly standard system (modification of Brodmann, 1912), it may be used as a simple anatomical reference by those readers unfamiliar with brain morphology. The labeled functional centers, however, must be viewed with a degree of skepticism.

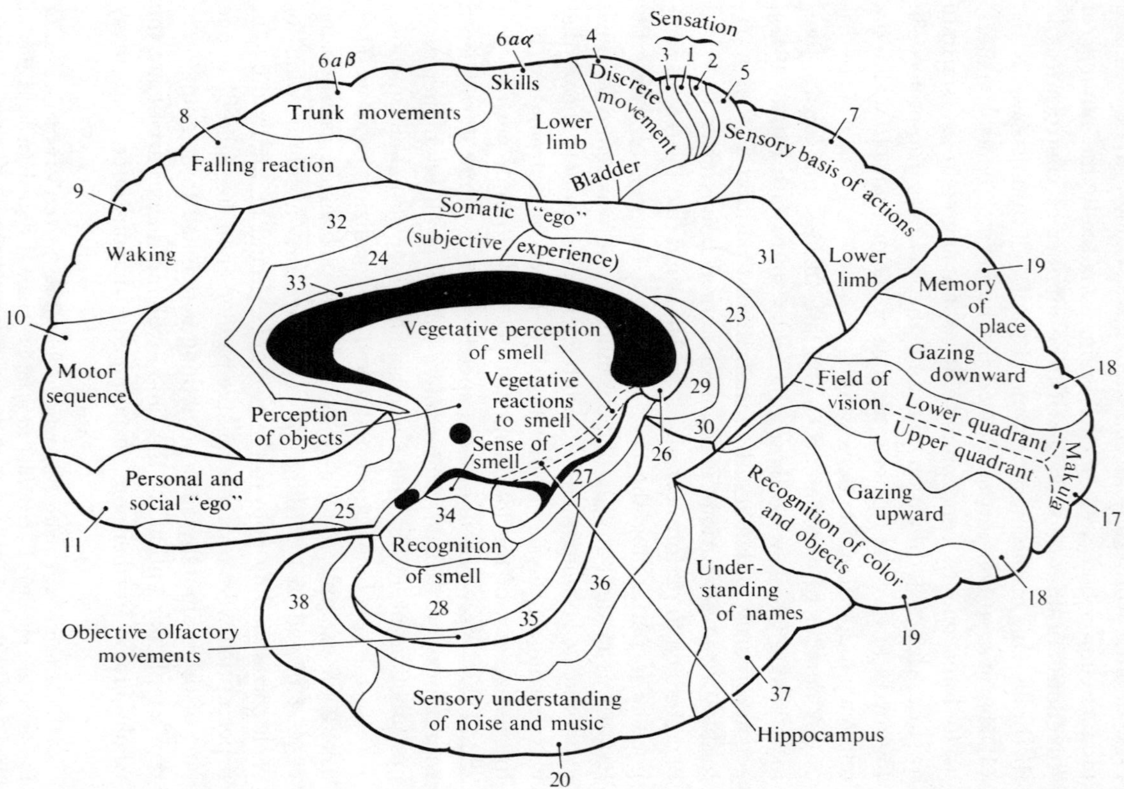

Figure 3.1. Kleist's (1934) localization chart. Reprinted with corrections from Higher Cortical Functions in Man by A.R.

sensory stimuli are perceived the way they "*ought*" to be, not necessarily the way they are physically presented. Thus, a coin at a slight angle is perceived as round instead of as an ellipse. All this, the Gestaltists argued, implied that perception is not a simple passive phenomenon but an active one involving, at the very least, many parts of the brain. Theories were presented explaining the mind as an "energy field," and Lashley himself developed the opinion (later significantly modified) that there was no differentiation of function in cortical tissue whatsoever. Needless to say, this position is as extreme on the part of the generalists (in the face of good ablation and stimulation evidence to the contrary) as was the earlier extreme localizationist stance.[4]

Recent research is moving toward a resolution of the problems, and, as one might have predicted, the truth seems to be somewhere in the middle.[5] In his consideration of the brain as an "engineering" problem, Gregory (1961) attempts to define what exactly is meant by localization of cerebral function. He concludes that the only possible meaning is that a structure is necessary, but not sufficient, to a given function. With that formulation we are sure few contemporary neuropsychologists or neurophysiologists would disagree. He gives the example of Broca's area (Brodmann's area 44) in the inferior convolution of frontal lobe on the dominant side. It has been well known since at least 1861 (Broca 1861) that if this area is destroyed a person is totally unable to speak. Accordingly this area has been known as the speech "center" since that time. Nobody doubts that Broca's area is necessary to speech, but it is just as certain that it is not in itself sufficient for speech. There are other necessary elements all the way from larynx through a myriad of intracerebral structures and paths. Why, then, Gregory argues, should Broca's area be singled out as the speech center? He gives an interesting example. If

[4] It is beyond the scope of this book to present a detailed history of the controversy or a summary of the evidence on both sides. If the reader is further interested in the development of the controversy, he is referred to the introduction in Luria's *Higher Cortical Functions in Man* (1966).

[5] An excellent summary of the presently evolving "middle position," although tending somewhat toward the generalist side, is to be found in Pribram's *Languages of the Brain: Experimental Paradoxes and Principles in Neuropsychology* (1971).

one removes a radio tube and a hum ensues are we to conclude that the purpose of that tube was to suppress the hum. Furthermore, the removal of several other tubes will produce exactly the same hum. So what can we conclude about the function of the first tube? Not very much!

We take a somewhat different approach from that of Gregory and those for whom any concept of localization whatsoever is anathema. As Gregory himself argues, the brain is compared more properly to a computer (whether digital or analogue is not resolved, Gregory 1961, pp. 226-230) than to a radio. Destruction of various critical areas involved in speech most certainly do produce a defect in speaking capabilities. But unlike the same hum that may be produced when any one of a number of radio tubes is removed, the deficit in speaking is distinctive and individually describable, depending on where the lesion is located. To be more specific (continuing with our speech example) a lesion affecting the nuclei (a group of neural cell bodies in the brainstem) of the nerves that control the muscles of speech and the vocal cords will produce a distinctive type of slurring technically called dysarthria. On the other hand, a lesion in the postcentral (parietal) area of the cortex (areas 1, 2, and 3 in Figure 3.1) may produce a totally different difficulty in speech arising from a difficulty in experiencing *feeling* of various positions of mouth and tongue and hence of being able to coordinate the subtle movements required for fluent speech. Consequently sounds may be jumbled together in a very distinctive manner technically called a kinesthetic aphasia. (An aphasia is any defect in either the understanding or expression of symbolic elements owing to a defect in higher brain or cortical areas.) In this case, although the symptoms are clearly motor (difficulty speaking), the real problem is a somaesthetic (feeling) deficit at the cortical level (Luria, 1966).

A third example is the classic lesion of Broca's area (area 44). If the area is totally destroyed a person simply cannot speak. If it is partially destroyed the difficulty is best described as one of being unable to order sequentially elements of speech into meaningful utterances. The patient knows clearly what he wants to say, but he appears to have lost the *idea* of how to say it. In linguistic terms (which we shall consider in detail later in this book) it appears that such a lesion produces a loss of the "master plan" of the surface structure of a language. Such a defect is called an expressive kinetic aphasia. Since speaking involves using verbal symbols such a defect certainly is technically an aphasia. But speaking is also a complex motor skill, and a loss of a previously attained motor skill owing to brain damage is technically called an apraxia. Some neurologists, therefore, have classified a lesion of Broca's area as an ideational apraxia. This means, simply, that the idea or master plan underlying the motor activity is lost.

From these examples we can see that three separate lesions produce three distinct and separately definable speech difficulties. Lesions in other areas produce other difficulties. If this is so, why should Broca's area be called the speech "center"? The answer is based on the value judgment that the area that appears to control the "master plan" of the surface structure of a language is more important than any of the secondary input or output structures that are related to other aspects of speech. To borrow an analogy from Aristotle, Broca's area (or, as we shall see, the Broca-arcuate fasciculus-Wernicke complex) might be regarded as the efficient cause of language whereas multiple other structures are simply necessary conditions. We acknowledge that much of this argument, on both sides, is semantic. We clearly recognize that for Broca's area to be functional an incredibly complex input is essential from many areas of the brain. But we also recognize that the evidence is overwhelming that there is something unique about this area of neural tissue that would not make it inappropriate to call it (and certain related structures) the "speech center."

This long digression is necessitated by the fact that we will occasionally use terms and arguments that some readers may consider too localizationist. We feel that not only is this appropriate but it is accurate as well, provided one keeps in mind that the brain is a highly integrated neural organ in which every part, to some extent at least, affects the function of every other part. But to deny that certain specific areas of the brain are more intimately connected with certain functions than other areas would be an absurdity in terms of our present knowledge.

Our position, we reiterate, is a middle one. On the complex

question of perception and memory, for example, we feel that the recent evidence is overwhelming that such phenomena cannot be totally explained in terms of simple one-to-one correspondence between, say, a point in the visual field and specifically localized cells in the visual cortex. It is true that Hubel and Wiesel (1962) have presented good experimental data that some such correspondence may indeed be operating during the initial stages of perception involving feature analysis (different lengths, edges, orientation of lines, etc.). But once a perception is "registered," so to speak, it appears to be stored *throughout* the visual cortex. Sperry et al. (1955) cross-hatched a sensory receiving area of the cortex and isolated the squares by placing mica strips deep in the surgical incisions. Kraft, Obrist, and Pribram (1960) and Stamm and Pribram (1961) produced multiple foci of epilepthiform discharges within a cortical sensory receiving area by injecting aluminum hydroxide cream. Such experiments, although they retard the learning of pattern discrimination, do not interfere with its execution after it has been learned. In reviewing these and similar experiments Pribram (1971) concludes: "These results make it clear that the effects of sensory input on brain tissue, the input information, must become distributed over the extent of the input system" (p. 122).

John, Herrington, and Sutton (1967) have produced evidence supporting the above conclusion from another type of experiment. They "tagged" an input into the visual system by presenting cats with stimuli that were differentiated by (among other variables) the frequency of the flickering light that illuminated them. The different frequencies were reflected by the neuroelectric brain activity that followed the imposed frequency. Among the results was the finding that identical wave forms were picked up in multiple related brain structures simultaneously.

Emphasizing the conclusion that perception is set down over relatively large areas of the brain Pribram (1971) notes:

> When 80 percent of the visual field is blinded by cortical removal, recognition is mediated by the remainder of the visual field; when the visual cortex is peppered with lesions, the part between the lesions functions so well that little difficulty is experienced in making discriminations; whether we view an object with one part of our retina or

> another, or whether we view it from one angle or another, we can still recognize the object. These are not the properties of ordinary photographic images—tear off 98 percent or even 80 percent of most photographs and try to identify them! (p. 124)

This may explain Lashley's original findings. He extirpated large portions of animal brains with astounding preservation of recognition. This undoubtedly was due to the fact that, if even a small part of the visual input system is spared, recognition is possible to some degree at least.

The reader must note carefully that although much evidence points to the fact that a total perception is stored almost everywhere within the pertinent sensory input system, there is *no* evidence that it is stored everywhere in the brain. In fact, the evidence is to the contrary. Although only a small amount of the visual input system is required for recognition, all recognition is lost if the system is *totally* extirpated. Thus the current evidence with respect to perception and recognition supports a position intermediate between the extreme localizationists and the extreme generalists. Perception of a sensory stimulus depends upon the pertinent sensory input system (localization) but appears to become distributed throughout that system, any part of which seems to register the whole (generalization). How this latter phenomenon is possible is a complicated question that we shall consider in chapter 4 when we discuss the holographic model for information storage.

We must now return to the question posed at the beginning of this chapter: Is it legitimate to consider the function and development of various subsystems of the brain apart from the brain as a whole? We feel, from the evidence presented in the foregoing discussion, and from that which will follow, that the answer is a definite "yes," although with the qualifications presented above. Indeed if we wish to gain some understanding of the evolution of hominid behavior, we feel that unquestionably the best approach is via a consideration of the function and development of cerebral subsystems.

Insofar as some anatomical structures (and presumably the functional adaptations they represent) can be demonstrated to occur successively on the phylogenetic scale, and insofar as certain anatomical structures within the brain can be shown to have some specificity of function, we shall take the liberty of referring to the evolution of multiple adaptive systems within the central nervous system for the purposes of this analysis. The reader is once again cautioned to bear in mind that all these systems are fully integrated in any given organism in which they are present. Obviously the most complex structures are not present in relatively primitive organisms. But when present, each system is related to all others via multiple neural pathways, and the whole functions quite harmoniously. Furthermore, we do not wish to give the impression that we are presenting a detailed account of the neurophysiology of higher cortical functions. Such a treatment is totally beyond the scope and intent of this book. We are dealing only with the major adaptive systems of the brain that bear directly either on hominid adaptation or on the adaptation of man's prehominid ancestors.

THE ASSOCIATION AREAS

With these caveats in mind let us turn to the evolution of the association areas in the brain. Each sense modality, sight (area 17), hearing (area 21), touch (areas 1, 2, 3), etc., has primary representation in the cerebral cortex that directly receives fibers carrying encoded messages from the periphery (body surface) according to that sense modality. There is likewise a primary focus of motor function in the cerebral cortex (area 4). These areas are called nuclear zones or primary cortical areas either for motor function or for the various sense modalities (Luria, 1966, pp. 58-60). They are clearly differentiated in terms of cellular structure (cytoarchitechtonics) in prosimians. With the evolution of the higher primates, one begins to notice the evolution of clearly differentiated "association areas" around each nuclear zone. These association areas represent the first cerebral adaptive mechanism we shall consider. Their purpose in general is to modify and modulate the function of the nuclear zones. Thus the association areas of the visual nuclear zone (areas 18 and 19) are responsible for the recognition of the object perceived. If the visual association areas are destroyed one of the effects is to obliterate recognition of objects although they are still "seen" (Luria, 1966, p. 133). The same is true of the auditory association areas (42, 22, 37, and adjacent). If the latter are extirpated, a sound is heard but not recognized specifically. An analogous relationship obtains between the primary somaesthetic (touch) area and its association areas (7 and adjacent). The association areas of the primary motor cortex include most of the frontal lobes (areas 6, 8, 9, 10, and adjacent); they are complex in function and involve the "voluntary" component of motor activity as well as laying down the patterns for such "skilled" motor sequences as walking, talking, and tying shoes (Luria, 1966, p. 222).

At this point let us consider only the gnostic or recognizing function of the sensory association areas. It is obvious that the ability to recognize precisely what a bit of sensory input signifies, and especially what relationships it has had with the organism in the past, is of great survival value. The sensory association areas allow the organism to store up information about an object in the external environment and to reactivate it when the object is once again presented as a stimulus. If a past association has been negative, the object will be avoided; if positive, the object may be approached.

THE VISUAL LIMBIC SYSTEM

But the total story is not so simple as that. The ability of an organism to associate positive or negative affect to a stimulus depends on the connections between the association area concerned and the limbic system (Geschwind, 1965). This is especially true of the visual association areas that have strong connections with the limbic system (hippocampus) in higher primates (Bonin and Bailey, 1947). Thus it appears that the gnostic effect of the association areas depends on the emotion originally associated with the stimulus. Indeed it is this pathway between visual association area and limbic centers that is responsible for most conditioned learning.

Viewed in another way, the system of connections between the visual association areas and the limbic system may have an adaptive function beyond that of being a basic mechanism of learning, recognition, and perhaps memory within the visual sensory modality. It is probably the mechanism by which sustained affect of any sort is invested in objects in the environment as they are presented to the visual cortex. It is this emotional investment in the environment that underlies motivated activity. In fact it may be this neural system function that is responsible for attachment behavior in infancy and that later provides an essential condition for group cooperation. Bowlby (1969) makes the point that attachment behavior in infants can, on many points, be compared with imprinting in birds. The original attachment to and "cathexis" of the mother, later modified and extended to other members of the same species by a mechanism to be described in chapter 5, probably is mediated via the visual-limbic connections, at least in part. If this is so, social living in lower animals and man requires this neural structure as a necessary condition, although the phenomenon may not totally be explained by it. The adaptive quality of social living has been stressed elsewhere (d'Aquili, 1972) This mechanism explains at least a part of the neural basis of socialization and of the cathexis both of external individuals and objects.

VISUAL-SOMAESTHETIC-MOTOR SYSTEMS

Another adaptive mechanism involves the increase of fibers connecting both the visual and somaesthetic (touch) association areas with the major motor association area among the Anthropoidea (Geschwind, 1965, pp. 612-615). In the first case, the increase of direct visual-motor connections makes possible increased skilled motor activity as a response to visual stimuli. This can be viewed as a preadaptation to tool using and tool making. The same thing, in general, can be said of the somaesthetic-motor connections. Since the somaesthetic association areas are not only concerned with recognition by touch but are also responsible for orienting things in space, it becomes obvious that the somaesthetic-motor connection "preadapted" animals to tool using, whatever the original adaptive mechanism may have been that resulted in the development of this neural fiber connection.

We have spoken of both these connections as "preadaptations" to tool using because they provided necessary conditions for tool using. Strictly speaking, there is no such thing as preadaptation. The actual adaptation for which they were selected was in all probability simply increased motor skill in response to the general environment. When the capacity to make tools finally evolved, these were clearly facilitating mechanisms. They were not, in other words, sufficient conditions in themselves for the development of tool using.

FRONTAL AND FRONTAL-VISUAL SYSTEMS

We now arrive at the adaptive mechanism that probably characterized our immediate prehominid ancestors and still characterizes Goodall's (1964) Tanzanian chimpanzees, i.e., the ability to make and use primitive tools.,

Tool using is a function of the ability to solve problems presented in the immediate environment by perceiving pertinent relationships within a perceptual field. This in turn required disattending irrelevant elements, and organizing relevant ones both spatially and temporally. Such an ability is a function of the frontal lobes and particularly of the anterior convexity (Luria, 1966, pp. 222-224; Pribram and Luria, 1973). Numerous experiments in animals and evidence from pathological lesions in man have demonstrated that the frontal lobes control sequential ordering of elements as well as attention to relevant elements and disattention to irrelevant ones. They have been shown to regulate goal-directed behavior. Le Gros Clark notes the progressive increase in size of the frontal lobes with the evolution of the Anthropoidea (Le Gros Clark, 1963). Bonin and Bailey (1947) have further demonstrated direct reciprocal connections between the anterior convexity of the frontal lobe and the visual association cortex in macaques. This connection is much more pronounced in chimpanzees and is undoubtedly responsible for their ability carefully to scan their visual fields and to so relate perceived elements as to be capable of solving problems presented within the same visual context. The solution of such problems involves the use and even the preparation of simple tools. Primitive tool using at the pongid level, therefore, had to await the development of fairly large frontal lobes together with, at least, their connection to the visual association cortex. It is probably at this level of development that we get the selection for tool users that we shall consider later as generating the beginnings of the hominid line.

FRONTAL-THALAMIC SYSTEMS

A further adaptive mechanism involves projections from the highly developed frontal lobes to the thalamus (perhaps ordering and filtering incoming stimuli) as well as projections to all sensory association areas. Luria (1966, p. 233) states that these connections synthesize information about the outside world with information about between "self" and "other." This can be seen as a "preadaptation" for the subjective reflection necessary for conscious thought. At any rate, it seems to be a requirement for setting up any sort of conscious relationship between subject and object essential to goal-directed behavior (such as problem solving).

THE "INFERIOR PARIETAL LOBULE" OR PARIETO-OCCIPITAL AREA

Let us now consider those cerebral adaptive mechanisms that seem to be peculiar to man. Earlier we mentioned that learning in animals depends on an intact connection between a specific sensory association area and the limbic system. Associations, furthermore, are always made between elements within the same sensory modality and form the basis of animal recognition of an object. Except in chimpanzees, there are no anatomical pathways allowing *direct* association between sensory modalities. Any such associations, then, are indirect for most animals, via the limbic system (e.g., instrumental or operant conditioning). As Geschwind (1965) points out, man has a cytoarchitechtonically[6] distinct area comprising an overlap of the visual, auditory, and somaesthetic association areas where *direct* cross-modal association is possible. It comprises the supramarginal and angular gyri and some adjacent areas (roughly areas 39 and 40

[6] Cytoarchitechtonics refers to the arrangements and organization of cells, in this case various kinds of neurons or nerve cells.

and their environs). Geschwind calls this area the "inferior parietal lobule." It is present in only rudimentary form in the chimpanzee and can be described as distinctive of man, at least functionally. The importance of this area cannot be overestimated. As Geschwind (1965), Holloway (1968a), and d'Aquili (1972) have all emphasized, the evolution of the dominant inferior parietal lobule in man does not mean that sensory-limbic learning has ceased to be important in man. It does mean that man's capacity to learn is not completely bound to the limbic system. Consequently, not only is a given percept "recognized" by the associations of that sense modality but also cross-modal transfer permits all the stored associations of the other nonlimbic sensory modalities to that percept. We can consider the analogy of several distinct computers, each programmed with different information about similar objects, being connected to each other and the information of each being made available to all. The possible permutations of associations would be staggering.

Here we find the basis for that process of categorization that we call conceptualization. For example, two visual percepts may generate more different visual associations than similar ones. If we use only visual associations, these percepts might not be classified together. But if we take into consideration all the possible associations in all sense modalities, it is probable that similarities frequently override dissimilarities, allowing large groups of objects to be classified together. In other words, cross-modal transfer permits the construction of classes of objects where formerly insufficient information was available in any one sensory storage system to serve as the basis for that classification. We now have an adaptive mechanism that is capable of generating classes of objects or, in more traditional language, *concepts.* The implication of this mechanism in the genesis of language is far reaching. It opens up the possibility for a referential system of communication independent of the immediate environment in which the communicators find themselves (demonstrative communication). It provides the basis for a system of communication using arbitrary symbols for classes of objects. However, such a communication system also required other neural mechanisms, which we shall describe shortly.

That this general area of the brain (inferior parietal lobule or parieto-occipital region) is intimately related to concept formation and categorization was suggested by clinical investigators as long ago as the early twenties (Goldstein and Gelb, 1924; Weigl, 1927; Conrad, 1932; and Zucker, 1934). For basically theoretical reasons Luria (1966) finds it hard to believe that this relatively small area of the brain can be responsible for such a high order of intellectual functioning as conceptualization. He, therefore, opposes himself to the above members of the "noetic" school, although he is forced to admit that at least some of the patients he studied did suffer from defects of abstraction and conceptualization after lesions in the parieto-occipital region. Apparently Luria was unaware of the work of Geschwind, cited above, which, we believe, tilts the balance in favor of the so-called noetic school. Luria does, however, give a masterful review of the logical-grammatical functions that are also impaired by lesions of this area (Luria 1966, pp. 154-158). Among these is the basic ability to organize external objects according to spatial antinomies (below-above, before-after, in-out, up-down, etc.). The importance and universality of antinomous relationships in "depth psychology," folklore, and anthropology will be considered in chapter 5. Furthermore, the ability to align one object in direct relation to another (even if not antinomously), such as with the possessive case (mother's brother) or the comparative degree of adjectives (A is bigger than B), is lost altogether with the destruction of this area. It is probable, therefore, that the neuroanatomic base of "binary opposition" (as posited by Durkheim, Hertz, Levi-Strauss, and others of the French structuralist tradition) may, at least in part, be found in this area of the brain. In addition, individuals with lesions of the parieto-occipital area (inferior parietal lobule) suffer difficulty transposing active to passive voice as well as determining the agent in either voice. This is suggestive, at least, of some difficulty making causal linkages. We shall return again to these critical issues in chapter 5.

It is interesting that Western investigators tend to emphasize the conceptual or categorical function of the inferior parietal lobule, whereas Soviet investigators emphasize its logical-grammatical functions. The discrepancy may in part be determined by theoretical perspective. However, since all subjects studied are human beings, their lesions are the result of natural disasters, and the precise locus of a lesion is rarely exactly the same in any two men; nor for that matter is the extent of damage. Since the functions we are considering are, for the most part, peculiar to man, one cannot make controlled lesions in animals to determine the extent and quality of function of this crucial region. Hence, one must piece together the available data from "natural" experiments. We contend that the empirical evidence clearly indicates that the inferior parietal lobule mediates *both* conceptual and logical-grammatical functions to a marked degree. On the whole then, it is certainly an area of utmost importance to the development of both reflective thought and language. As Le Gros Clark (1963) notes, the enlargement of the parietal lobe was very great in the evolution of the hominid brain, actually "forcing" most of the visual cortex medially. The only area of comparable development in hominid cerebral evolution is the anterior region of the frontal lobes.

Up to this point, we have been concerned with the tremendously important function of conceptualization and the basic logical-grammatical categories that the dominant inferior parietal lobule seems to generate. Although we cannot overestimate these functions in the development of language and culture, this structure may have a more primitive adaptive value that appears to be present from birth. Bower (1971) conducted a series of experiments concerning how infants learn to associate touch and vision in perceiving objects. His rather startling conclusion is that infants do not associate the different sense modalities at all (in the usual sense of "associate"). The association appears built in from birth. The traditional theory holds that the child associates characteristics of an object presented through different sense modalities by experiencing simultaneous, or nearly simultaneous, sensations through the different modalities. Thus a child learns that a block is hard by simultaneously seeing and feeling it. Although this is undoubtedly true for complex object discrimination at a later age, the awareness that gross visual experiences have gross tactile consequences seems present from birth, without the actual experience of those tactile consequences (Bower, 1971). His conclusion is pertinent: "We can only conclude that in man there is a primitive unity of the senses, with visual variables specifying tactile consequences, and that this primitive unity is built into the structure of the human nervous system" (p. 32). We contend that this is a simple and very basic example of an inherited structure governing a model of reality. We shall consider such inherited structures in detail in chapter 5. We include this example here, however, because it has such obvious evolutionary significance. As Bower points out (p. 38), other primates do not appear to have this inborn ability and probably learn to associate qualities of an object via the traditional mechanism outlined above. The only trouble with this latter method of learning is that the initial trial of associating qualities might be the final one if an animal were exposed to a life-threatening situation.

So important for survival was this "primitive unity of the senses" that this inherent function was probably strongly selected for, thus firmly establishing the inferior parietal lobule in hominid evolution. One can look upon the development of "cross-modal" transfer of complex input, concept formation, and basic logical-grammatical categorization as derivative functions following increasing development of the area phylogenetically and increasing myelinization[7] ontogenetically.

NEURAL MECHANISMS UNDERLYING SPEECH

We now turn to the evolution of a mechanism essential to the development of speech. Geschwind refers to its anatomical locus as the speech center. By this he means the verbal auditory association area (Wernicke's area or area 22), the motor speech area (Broca's area or area 44), and their direct connection via the arcuate fasciculus. This latter fiber tract is distinctively developed in modern man. Indeed there is some question whether it is present in the chimpanzee at all (Bonin and Bailey, 1947). In any case this connection is well established only in man. The significance of such a direct connection can be seen in the findings of Cooper et al. (1958). They

[7] Myelinization is the process by which a fatty sheath is laid down around certain nerve fibers, increasing their conductivity and efficiency. The process continues for several years after birth in certain areas of the brain.

proposed that phonemic identification of a linguistic signal is decided by the listener in terms of the articulatory controls by which he would repeat the signal when acting as a speaker. Thus phonemic identification requires both the auditory association area and the motor speech area to be directly connected. It is interesting that a lesion of the arcuate fasciculus connecting the auditory association and motor speech areas results in an inability to repeat words (Geschwind, 1965). What the patient does say is numerous associations to the word given him. Rarely, he may eventually say the word itself, but it is obvious that it comes out as only one of multiple associations. This is interpreted to mean that every verbal input, once it reaches the auditory (verbal) association area (Wernicke's area), follows two paths: one directly to the motor speech area, there to result in its automatic utterance (unless it is inhibited by other mechanisms), and one to the inferior parietal lobule where multiple associations take place (how these multiple associations find their way out as speech will be considered below). The command to repeat certain words essentially "turns off" the inferior parietal lobule or, more properly, temporarily disconnects it from the speech center, thus allowing automatic functioning of the speech center. This interpretation is confirmed by the fact that if the inferior parietal lobule is destroyed the patient exhibits echolalia, i.e., he blindly repeats whatever is said to him (Geschwind, 1965, p. 632). Certain lesions of the frontal lobe also produce echolalia. Why this should be so will become clearer when we consider the mechanisms by which the multiple associations generated by the inferior parietal lobule are transformed into actual speech. At any rate, it is probably frontal influences that inhibit the simple reproduction of whatever is heard in ordinary conversation, thus allowing its repetition as "internal language" with consequent phonemic identification. It is tempting to see the evolution of the speech center as originally resulting in a strong tendency towards the reproduction of sounds emitted by others. Such a mechanism would account for fairly rapid group agreement on sound clusters for reference.

Although the evolution of the speech center offers a mechanism for phonemic identification as well as for social concord on sounds to be used in language, we have not as yet considered the mechanism for transferring the multiple associations, incipient concepts, and logical-grammatical relationships generated by the inferior parietal lobule into actual spoken language.

INTEGRATION OF THE FRONTAL, INFERIOR PARIETAL, AND SPEECH SYSTEMS

This brings us to the last cerebral adaptive mechanism we shall consider in this chapter. First of all, it is known that fibers from the inferior parietal lobule project to prefrontal areas (Luria, 1966, p. 230). Second, there are also connections between the motor speech area and the prefrontal region. Third, although a lesion of the motor speech area produces total inability to speak but leaves full comprehension, a lesion of the prefrontal area may produce confused associations of words or else "telegraphic" style. This latter is characterized by the use of nouns primarily and by the apparent inability to predicate. Furthermore, some lesions in the prefrontal area cause what appears to be a marked intellectual deficit (Luria, 1966, pp. 283-286; Pribram and Luria, 1973).

The best interpretation of these data is that the ordering of associations takes place in the prefrontal area in keeping with the general frontal function of sequential ordering and goal-directed operations. The input would come from fibers originating in the inferior parietal lobule. Such ordering probably conforms to basic logic and does not represent the actual sequence of the spoken language. Thus the frontal lobes "make sense" out of the multiple associations originating in the inferior parietal lobule. This material is then fed into the motor speech area, which then specifically orders it according to the spoken sequence of the particular language. Thus the speech center, and in particular the motor speech area, appears to contain not only the mechanism of phonemic identification but also the whole code for the surface structure of the language. Now we can see why a prefrontal lesion generates echolalia. By destroying the arc of connections between the inferior parietal lobule and the motor speech area via the prefrontal area, it denies the input into this area just as surely as when the inferior parietal lobule is itself destroyed. Under these circumstances, the speech center is released from inhibition and generates automatic repetition of what is heard.

It can be seen that the progressive development of the frontal lobes, so essential to more efficient problem solving and tool using, provided an essential "preadaptation" to the development of language. Without it, the multiple associations of the inferior parietal lobule could never be unscrambled and fed into the speech center to produce language. A direct connection, therefore, can be traced between adaptation to tool use and the development of language through the progressive enlargement and differentiation of the frontal lobes. It must be admitted, however, that the other two neural mechanisms essential to language production (the inferior parietal lobule and the arcuate fasciculus connecting the auditory association and motor speech areas) developed separately from any demonstrable relationship to tool using. Once they had developed, however, and once they had been integrated with the frontal system, man developed his most universally adaptive behavioral trait—language. With language, the basic primate social group had a communication system free from the exigencies of demonstrative reference and hence free from the stimuli of immediate environment. With this development, "culture" blossomed and with it the ability to adapt to almost any environment. The precise relationships of social grouping, language, and nonlinguistic social behavior have been considered in detail elsewhere (d'Aquili, 1972).

"Dominance" and Hemispheric Asymmetry of Function

We cannot move on to a consideration of the evolution of systems without first briefly discussing the question of hemispheric dominance. This is an issue that has been of interest in the neurological sciences since the 1860s (Penfield and Roberts, 1959) and one that attracted the attention of anthropologists as long ago as the turn of the century (Hertz, 1909). By dominance is meant that one cerebral hemisphere either totally, or at least primarily, controls significant functions of an organism. The preponderant concern for dominance has been in the cerebral control of speech and handedness. Thus the "dominant" hemisphere is that hemisphere that controls speech. This concern has recently been extended to cover abstractive ability, logical functions, and the perception of the qualitative relationships underlying mathematics.

What does this mean in terms of the gross neural structures we have been discussing above? It means that, although the gross structures are bilateral, only those on the "dominant" side control particular functions being considered. Therefore, lesions on the dominant side may produce profound impairment of one or several of these functions whereas similar lesions on the opposite ("nondominant," "minor") side leave these functions intact. These discoveries have led in the past to a number of questions about the functional purpose of the minor hemisphere and whether it controls any higher functions that are not similarly controlled by the dominant hemisphere as well.

Evidence obtained by Sperry et al. (1969), Nebes (1971), Nebes and Sperry (1971), and Nebes (1972) from patients with lesions destroying the connections between the hemispheres indicate that the nondominant hemisphere appears to have relatively specific functions of its own. In particular, these pioneering studies were able to demonstrate a superiority of the minor hemisphere for the perception of gestalt, or synthetic, relations. Other work has shown that the minor lobe may be intimately related to "creative" and "intuitive" operations, as opposed to the logical, abstract, and verbal processes controlled by the dominant hemisphere.

The findings of Sperry and his associates has led to a rash of studies directed at isolating hemisphere-specific functions. Virtually all these studies tend to support Sperry's findings. Without going into the technicalities of research design used, and recognizing the extremely gross and partial nature of the findings thus far, we will present in Table 3.1 the hemisphere-specific functions isolated to date. The reader must be careful to realize that the functions are grouped according to the assumption of left hemisphere dominance for langauge (the statistically usual case). Also, the categories listed

Left Hemisphere	*Right Hemisphere*
Auditory: Speech sounds. Processing consonant-vowel syllables. Symbol translation and analysis. Recall of auditory images of visual objects.	*Auditory:* Melody recognition. Vocal nonspeech sounds (coughing, laughing, crying). Steady-state vowels.
Visual: Perceptual recognition of conceptual similarity.	*Visual:* Drawing, building models from a picture. Space relations in two- and three-dimensional space. Perceptual recognition of identity.
Tactile and Motor: Contralateral control of motor functions. Fine hand motor skills. Hand gestures during speech.	*Tactile and Motor:* Contralateral control of motor functions. Tactile pattern perception.
Conceptual: "Logical-analytical" operations on perceptual material. Temporal sequential ordering.	*Conceptual:* "Gestalt-synthetic" operations on perceptual material.

Table 3.1: Lateral Asymmetry of Function in the Human Brain

for each column should not be thought of as mutually exclusive. These functions have been summarized from Kimura (1973), Bogen et al. (1972), Shankweiler and Studdert-Kennedy (1967), and Levy (1972). For a more speculative list of functions, the reader is referred to Ornstein (1972, p. 67).

In considering the emerging data on hemispheric asymmetry of function, some authorities have been led to suggest that the human brain may, in actuality, house two minds. On a much less dramatic scale, Levy (1972), after intimate involvement in "split-brain" research,[8] has said:

[8] "Split-brain" research refers to research carried out on humans and other animals after the major connective fibers integrating the two halves of the brain have been surgically sectioned.

> . . . I was led to the idea that perhaps there is an intrinsic antagonism between the modes of information processing of the two hemispheres. Perhaps a hemisphere which is capable of expressing itself in language does not merely have the capability of symbolic-analytic reasoning, but is, in fact, constrained to use such reasoning. Such a hemisphere thinks in terms of symbolic and not visual relationships. It tends to look at stimuli analytically, as if it intended to describe what it sees. If true, this idea provides a basis for understanding why in man, but in no other animal, there is such a profound functional differentiation of the two half brains. Once the ancestral hominids acquired the capacity for language, there would be an obvious adaptive advantage for that capacity to be confined to a single hemisphere, leaving the other free to carry on the perceptive Gestalt functions. Had both sides of the brain possessed language, there would have been a loss in visualization abilities. The old adage, "He can't see the forest for the trees" possibly expresses most succinctly the proposed antagonism. Perhaps a language-competent hemisphere is simply incapable of seeing forests. (pp. 171-172)

Relative to this, Professor Colwyn Trevarthen (1962a, 1969) of the University of Edinburgh has suggested that the hemispheres of animals may operate in accordance with the same principle of reciprocal inhibition that Sherrington discovered long ago for the simple reflex. That is, Trevarthen suggested that in order for the animal to respond effectively to a stimulus presented to its right side, its left hemisphere has to be activated and its right hemisphere reciprocally deactivated. The opposite would occur for a stimulus presented on the left side of the animal. In another paper Trevarthen (1962b) further suggested that in order for learning to occur in "split-brained" monkeys, attention must be directed to one or the other lobe (not both). The lobe to which attention has been focused would do the learning while the other lobe does not record the information. Following the work of Trevarthen, Levy (personal communication) has tentatively suggested that the opposed or antagonistic functions of the respective hemispheres in the normal brain may be alternately directed to the solution of any task by a process of rapid reciprocal inhibition across the corpus callosum.

The topic of hemispheric "dominance" is, for anthropologists, important to the issue of the evolution of language and handedness.

There are two competing hypotheses worthy of note which might account for the evolution of asymmetry of function, both of which have sound reasoning and some evidence to support them. We will call these the *specialization hypothesis* and the *preadaptation hypothesis* respectively. We will briefly summarize their central arguments.

SPECIALIZATION HYPOTHESIS

This hypothesis states that the extreme hemispheric asymmetry characteristic of contemporary man, and probably characteristic of *Australopithecus*,[9] was the result of specialization requisite to the development of speech. If we consider the hemisphere-specific functions summarized in Table 3.1 for the right or minor lobe, the view would hold that the minor hemisphere did not so much evolve dominance in these areas as that the major hemisphere *lost* equal facility in them as it specialized in the logical, abstractive, and linguistic functions. In other words, the functions exhibited by the minor hemisphere may represent the "primitive" or basic functions of the entire, generalized brain of the prehominids. These functions being requisite to continued survival, they became localized in the minor lobe, thus leaving the dominant lobe free to specialize. This position assumes, of course, that man is unique in exhibiting asymmetry of function—that monkeys and apes, for instance, have completely symmetrical brains.

Luria (1966) suggests that comparison and contrast in cognition (and language) developed out of the basic parietal lobe function of spatial orientation (up-down, in-out, etc.). From the work of Sperry and Nebes it would appear that important and basic spatial perceptions are governed primarily by the minor hemisphere. If this hypothesis is correct, this function in the major hemisphere must have been sacrificed to the evolution of cognitive processes that, nevertheless, ultimately were based on them.

[9] Dart (1949) evaluated a sample of kills made by *Australopithecus* and reached the conclusion that the weapons used in the kills had been wielded primarily by a right-handed population.

PREADAPTATION HYPOTHESIS

The second, and more controversial, hypothesis states that there existed in the prehominids a basic asymmetry of function that operated, perhaps in the manner suggested by Trevarthen and Levy above, to direct opposing functions to problems or tasks of survival. This alternation of asymmetric "talents" set down a preadaptive foundation for later modification in the acquisition of speech. As hominids began to specialize in vocal communication, the hemisphere which handled analytic, logical sequencing also provided the matrix from which the neural structure developed to mediate the behavior. The effect of linguistic specialization, in this view, was to broaden an already existing separation of function. The hypothesis would deny that man is unique in having asymmetry of function—that asymmetry of function is a fundamental principle of structure and function in the generalized vertebrate central nervous system, especially in the nervous systems of the primates.

To date, no clear evidence for hemispheric asymmetry has been documented for primates, or for any mammals as far as we know. Some evidence exists that attempts to cause measurable behavioral and conceptual deficits by unilateral lesion have been largely ineffective. Much research has centered on tests for paw preference, which, it is assumed, would be some measure of "dominance" as in man. These studies, however, have given contradictory results. Warren (1953), for example, found strong hand preference in rhesus monkeys, whereas Lehman (1970) found a continuous range of hand preference in the same species from extreme right-handedness, through degrees of ambidexterity, to extreme left-handedness.

It is interesting in this regard that Nottebohm (1971, 1972) found neural lateralization of vocal function in the European songbird, *Fringilla coelebs*. The lateralization, incidentally, was in favor of the left lobe. The question naturally arises as to whether this commonality between man and bird is the result of widespread preadaption for hemispheric asymmetry or is merely an example of parallel development.

It should now be evident to the reader that the issue of the evolutionary significance of cerebral dominance is a complex one indeed. We would suggest, however, that the issue might easily be resolved by a comprehensive and concerted study of perceptual and cognitive functioning in split-brained primates—with a particular emphasis upon the chimpanzee, which we now know to be our closest living relative. If the chimpanzee proves to have no hemispheric asymmetry in these functions, that fact would go far in supporting the specialization hypothesis. If, on the other hand, the chimpanzee and/or other primates prove to have some degree of asymmetry, such findings would tend to support the preadaptation hypothesis.

HANDEDNESS

We have purposely avoided a lengthy discussion of the relationship between handedness and speech dominance. We have done so because it would make an already difficult subject complex beyond the overall intent of this book. Yet, it would be well to mention that the traditional assumption of inevitable contralateral (opposite-side) control of hand preference by the speech hemisphere in man has come under recent serious question. Jerre Levy and her associates (Levy and Mandel, 1972; Levy and Nagylaki, 1972; Levy and Laughlin, 1974) are, at present, attempting to clarify the relationships between hemispheric asymmetry, degree of asymmetry, specific perceptual and cognitive functions, speech, and handedness.

As the reader is at this point no doubt aware, the area of research so briefly discussed in this section is potentially pregnant with implications for the study of the neural substrate of human behavior. For example, there is increasing evidence that at least some of the perceptual and motor functions isolated in split-brain research are sex-linked (Kimura, 1973; Levy, 1972). Also, evidence is accumulating that suggests that differential reliance upon particular cerebral functions may be due in part to different ethnic group-specific patterns in ontogeny (Bogen et al, 1972). Finally, Barbara Lex (1973) has suggested that the widespread incidence of trance states, use of chants and mantras, and use of hallucinogenic drugs in ritual may be explained by their effect in "turning off" the logical-analytic functions of the dominant hemisphere and allowing access to the functions of the nondominant hemisphere.

Since we are here concerned with the evolution of those neural functions involved in the structural basis of sociocultural phenomena, we have been discussing primarily the functions of the dominant hemisphere (i.e., language, abstract problem solving, logic, etc.). And because we have discussed these functions in an evolutionary context, we have called them cerebral adaptive mechanisms. Let us see if we can more specifically fit the development of these cerebral adaptive mechanisms into the matrix of hominid evolution.

The Evolution of Cerebral Adaptive Mechanisms

Le Gros Clark, in *The Fossil Evidence for Human Evolution* (1964, p. 3), noted that the human brain is little more than that "magnified model of the brain of an anthropoid ape." In *The Antecedents of Man* (1963), the same author remarked that there are no morphologically distinct areas of the human brain that are not represented, to some degree at least, in the brain of the ape. If this is true, one would expect that, under certain conditions at least, apes could exhibit behavior heretofore considered distinctively human, such as tool making, cross-modal transfer, and arbitrary (noniconic) use of symbols. Recently all these behaviors have either been observed in field studies of chimpanzees or else elicited under controlled laboratory conditions. Goodall's observations (1964) of obvious tool using among Tanzanian chimpanzees are by now well known. What is significant here is that this activity is spontaneous and hence reflects cerebral mechanisms not only *capable* of generating tool using but sufficiently developed to result in such behavior spontaneously in the wild. This alone suggests a very high degree of development of frontal lobe mechanisms in the chimpanzee. Indeed, in terms of strictly frontal lobe functional capability, *Homo sapiens* may not be significantly removed from the chimp. Brummelkamp and Offringa (1940) have shown the relative size of the frontal lobe in man and apes to be the same. Of course insofar as the frontal lobe of *Homo sapiens* sequentially orders the functioning of parts of the brain that are only rudimentary in the ape, it can be said to function at a "higher" level than that of the ape. But this is due to the minimal development of other areas, not of the frontal lobe itself. This observation leaves us with the somewhat startling conclusion that perhaps the chimpanzee already possesses a capacity for frontal lobe function almost equal to that of *Homo sapiens.* It is rather the primitive nature of the inferior parietal lobule and the speech center that limits the potential of the frontal lobe and creates the intellectual gap between the ape and modern man.

But let us return to the point that all, or almost all, elements of cerebral morphology present in man are present at least in primitive form in the ape. As noted in the previous section, the inferior parietal lobule, so essential to "concept" formation derived from cross-modal transfer, is present only in very rudimentary form in the chimpanzee. Nevertheless, it is present. One would therefore expect that some degree of cross-modal transfer and symbolization would be elicitable in the chimp. Although such behavior has never been observed in the field, the Gardiners (1969) among others (see Fleming, 1974) have presented evidence suggestive of cross-modal transfer and non-iconic symbol usage in the chimpanzee. Preliminary reports of the Yerkes project (*Medical World News,* 1971) indicated that cross-modal transfer is definitely possible for the chimp under carefully controlled laboratory conditions. Such behavior is never spontaneous, however. This probably indicates that although the neural capability is developing, it has not yet become a sufficiently developed cerebral system to operate within a natural environment.

With reference to speech, it must be admitted that nobody has been able to produce even the most elementary examples of true speech in an ape. This possibly reflects something that *is* peculiar to man. i.e., functional connection of the auditory (verbal) association area with Broca's area, and the functional connections of this speech center both with the inferior parietal lobule and with the anterior convexity of the frontal lobe.

Taken together, however, the neural reorganization that Holloway (1966) postulates for the australopithecines, and that we alluded to in chapter 2, seems already to have begun to take place in man's prehominid ancestor, assuming the chimpanzee approximates that creature. Thus the present-day chimpanzee can be seen as already on the threshold of a "cultural" explosion, simulating man's prehominid ancestor. It remained only for the inferior parietal lobule to expand in order to have an animal fully capable of spontaneous cross-modal transfer and hence of non-iconic symbolization and "culture." What we mean by the term "culture" in the present discussion is the capacity for generating socially adaptive, cumulative conceptual models that do not necessarily require environmental reinforcement at each generation. Note in this context that the connections of the speech center need not have evolved such cumulative models. Once language became the vehicle for arbitrary symbolization, however, one would expect a rapid spurt in the complexity of model formation and culture. Nevertheless, culture is conceivable without verbal symbols and there is no reason, *a priori*, why the first "culture bearers" must have evolved speech. They must, however, have been able to communicate symbolically, at least to a limited extent, which would have involved both the inferior parietal lobule and the frontal lobe. Let us consider how this evolution from pongid to hominid is likely to have taken place.

Contrary to the view of some authors (Robinson, 1962; von Koenigswald, 1962), it is highly unlikely that the hominid line split off from its pongid-like ancestors on the basis of the evolution of bipedalism. Actually, if evolution is really controlled by successful adaptation, it is much more likely that the original change was behavioral, accompanied by reproductive isolation. Morphological changes responsible for bipedalism (and facial and dental changes as well) would have been subsequent to the original behavioral change. Let us give a hypothetical example that bears directly upon the evolution of higher cortical functions. We will presume, as many authors do, that the ancestral prehominid markedly resembled the present-day chimpanzee. Now, on the basis of Goodall's observations (1964) noted above, we know that some populations of chimpanzees are tool users, albeit to a limited extent. This already presupposes the evolution of the frontal mechanisms we have considered in the previous section. Now let us suppose that environmental changes (such as progressive desiccation in Africa during a period of the Early Pleistocene) isolated forest areas, thus effectively isolating populations of prehominids; and let us suppose, furthermore, that such changes drastically limited a previously plentiful food supply (fruit trees) by encroaching upon available forest. Under such conditions, those animals most adept at tool using (i.e., those with the neural mechanisms permitting such behavior) would be most likely to survive by employing tools, however primitive, in getting food. In a reproductively isolated setting natural selection would rapidly favor the tool users. What was once an incidental accompaniment to life (tool using) became central to survival. Evolution consequently favored the development of those neural mechanisms that improved the efficiency of tool using. Obviously, any gross morphological changes that tended in the same direction would also be favored, such as bipedalism, which effectively resulted in freeing the hands for use. This view maintains therefore, that in the prehominid population tool using was the significant variable that was selected for by environmental change. It became the critical adaptive pattern that was to distinguish the hominid line.

An adequate analysis of hominid adaptation thus requires a consideration of the development of the central nervous system insofar as it generated an increased ability both to make and to use tools. Such an analysis must necessarily include the ultimate evolution of symbolic communication, since it freed the tradition of tool making, and especially that of tool using, from the constraints of immediate imitative learning. Thus it is probably in successful tool using that we find the adaptation that was ultimately responsible for the evolution of all higher cortical functions. Furthermore, few would deny that technology in one form or another represents one of the primary adaptive elements of human society even today. As was pointed out earlier, the evolution of this behavior can be explained on the basis of a neural reorganization involving primarily the development of the essential neural connections that we considered in the previous section. Such reorganization does not require any significant increase in brain size over that of the pongid brain. We have already shown that the chimpanzee has most of the essential connections required for hominid-like adaptation, although only the frontal and frontal-visual mechanisms are sufficiently developed to operate spontaneously within a natural environment. Consequently, there is absolutely no reason to deny cumulative modeling, *a priori,* to the australopithecines, on the basis of a small cranial capacity alone. Indeed examination of the endocranial casts of all the australopithecines, and in particular the Taung skull (Le Gros Clark, 1964, p. 135), indicated an appreciable increase of the inferior parietal area relative to other areas of the brain. The development of this area is certainly significantly greater in the australopithecine brain than in the pongid brain. It is only the rudimentary development of this structure in the chimpanzee that prevents him from spontaneous cross-modal transfer and non-iconic representation.

The development of this area in *Australopithecus,* combined with highly developed frontal functioning (already attained at the pongid level), is in itself sufficient to make highly probable the statement that the australopithecines were "culture bearers." The association of australopithecine fossils with lithic industries should provide no problem to anthropologists except to those who adhere to the rigid association of culture with brain size rather than with a certain pattern of neural organization. This is not to say that the marked increase in cranial capacity during the Pleistocene did not have behavioral correlates. Indeed language probably developed with this expansion. Most of the change probably involved increasing efficiency and redundancy, rather than the development of basic capacities.

This brings us to the question of whether the australopithecines possessed speech. That speech is not essential for cumulative modeling, or "culture," we have already indicated. The problem, however, is one for which there can never be a definitive answer, since we will never be able to dissect an australopithecine brain. According to the thesis presented in this book, it is possible that the neural reorganization accompanying the gradual pongid-hominid transition did include the basic connections required for speech as outlined in the previous section. Whether these connections, if they existed, were sufficiently developed to have been spontaneously functional is a moot question. Neither the temporal area involving the auditory association areas nor the inferior frontal convolution containing the motor speech area in *Homo sapiens* seems markedly developed in *Australopithecus* over homologous areas of pongid endocasts. This is in marked contrast to the obvious parietal development that *Austra-*

lopithecus shows. It may very well be that the areas and connections associated with speech, while present in rudimentary form, were not sufficiently developed to be functional—a situation analogous to the function (or rather lack of it) of the inferior parietal lobule of the chimpanzee under natural conditions. Hence we would be tempted to say that *Australopithecus* did not have speech, although it is possible that the requisite areas and connections were just sufficiently developed to be functional but not sufficiently developed to leave gross changes noticeable on the endocranial cast. If this latter was the case, it is probable that the speech mechanisms were not very efficient. Consequently, one would expect that if language was present at all, it would not have resembled the fluent speech of *Homo sapiens.*

It is precisely the evolution of speech, or at least a reasonably fluent speech, that seems to have most markedly characterized the development of the genus *Homo*. Early descriptions of *Homo erectus* (Black, 1933; Shellshear and Smith, 1934), whatever may have been their shortcomings, all agree that these fossils are characterized by clear and obvious development of the second temporal convolution, the inferior frontal convolution, as well as the inferior parietal area. The prominence of the first two structures strongly suggest that *Homo erectus* possessed speech. The relative increase of the inferior parietal lobule over that of *Australopithecus* indicates an increased facility for cross-modal transfer and "concept formation." But it is the development of the areas associated with speech that most clearly characterizes *Homo erectus* with respect to his most important behavioral adaptation. The evolution of speech must certainly have marked the beginning of a period of rapid culture development over that of the relatively nonverbal australopithecine culture bearers (who nevertheless probably had some form of noniconic symbolization as a basis for communication). It seems hardly coincidental that the use of fire and the precocious development of advanced stone tool industries characterized the advent of *Homo erectus.*

Kotchetkova (1970) notes with respect to the brain of *Homo erectus*, ". . . all investigators have noted that certain sections show more intensive growth. We marked them as the lower parietal and lower frontal eminences. . . . Sections of the lower parietal lobule are known to effect a very complicated analysis of the signals that reflect various properties of objects placed in the immediate vicinity of the subject. . . ." Thus, the main difference between *H. erectus* and the Australopithecinae lay in the perfecting of those mechanisms of the brain that furthered the development of the inferior parietal area, the areas associated with speech, and some heightening of the frontal lobes. Except for speech, it is questionable whether these changes in *Homo erectus* were accompanied by any qualitative change in the cerebral adaptive mechanisms with which these areas are concerned. More probably, the expansion represents redundancy in biologically important systems.

In this chapter we have considered the various mechanisms underlying higher cortical functions together with their adaptive value in an evolutionary context. Morphological and functional considerations have necessarily been of a gross nature. When we state that recognition of an object presented in a given sensory modality is stored in the association areas of that modality, what do we mean in actual physical terms? When we state that the inferior parietal lobule is involved in concept formation, or that man relates concepts to each other in various specific ways, do these statements have any neuroanatomical meaning? Obviously they must, unless we wish to consider man's "mind" as an epiphenomenal entity unrelated to neural functioning—a position that we categorically disavow. It is clear that we are much more ignorant in these areas than knowledgable. However, we feel strongly that legitimate inferences may be made concerning some of these specific areas of neural activity based on what is already known. It is the complex structure of human thought and its probable neuroanatomical bases that will be the topic of the next two chapters.

References

d'Aquili, E. 1972. *The Biopsychological Determinants of Culture*. Reading: Addison-Wesley.

Black, D. 1933. "On the Endocranial Cast of the Adolescent *Sinanthropus* Skull." *Proceedings of the Royal Society, London*, B, 112:263.

Bogen, J. E., R. DeZure, W. D. Tenhouten, J. F. Marsh. 1972. "The Other Side of the Brain IV. The A/P Ratio." *Bulletin of the Los Angeles Neurological Societies,* 37(2):49-61.

Bonin, G. von, and P. Bailey. 1947. *The Neocortex of Macaca Mulatta.* Urbana: University of Illinois Press.

Bower, T. G. R. 1971. "The Object in the World of the Infant." *Scientific American,* October, p. 30.

Bowlby, J. 1969. *Attachment and Loss,* Vol 1. New York: Basic Books.

Broca, P. 1861. Remarques sur la siège de la faculté du langage articulé, suivies d'une observation d'aphemie (perte de la parole)." *Bulletins de la Société Anatomique de Paris,* Tome VI, 36:330-357.

Brodmann, K. 1912. "Neurere Ergebnisse über die vergleichende histologische Lokalisation der Grosshirnrinde mit besonderer Berücksichtigung des Stirnhirns." *Anatomischer Anzeiger*:211.

Brummelkamp, R., and J. Offringa. 1940. "The Relative Growth of the Frontal Brain during Human Onto- and Phylogenesis." *Acta Neerlandica Morphologica,* 3:202-208.

Conrad, K. 1932. "Versuch einer psychologischen Analyse des Pareitalsyndromes." *Monatsschrift für Psychiatrie und Neurologie,* 84.

Cooper, F. S., A. M. Lieberman, K. S. Harris, and P. M. Grubb. 1958. "Some Input-Output Relations Observed in Experiments in the Perception of Speech." *Proceedings of the Second International Congress on Cybernetics,* Mamur, Belgium.

Crosby, E. C., and T. Humphrey. 1941. "Studies of the Vertebrate Telecephalon." II. *Journal of Comparative Neurology,* 74:309.

Daitz, H. 1953. "Note of the Fibre Content of the Fornix System in Man." *Brain,* 76:509-512.

Dart, R. A. 1949. "The Predatory Implemental Technique of Australopithecus." *American Journal Physical Anthropology,* 7:1-38.

Fleming, J. D. 1974. "Field Report: The State of the Apes." *Psychology Today,* 7(8):31 ff.

Gardiner, R. A., and G. T. Gardiner. 1969. "Teaching Sign Language to a Chimpanzee." *Science,* Vol. 169, No. 3894, pp. 664-672.

Geschwind, N. 1965. "Disconnexion Syndromes in Animals and Man." *Brain.* 88:237-294; 585-644.

Goldstein, K., and A. Gelb. 1924. "Uber Farbennamenamnesie." *Psychologische Forschung.* 6.

Goodall, J. 1964. "Tool Using and Aimed Throwing in a Community of Free-Living Chimpanzees." *Nature,* 201:1264-1266.

Gregory, R. L. 1961. "The Brain as an Engineering Problem." In W. H. Thorpe and O. L. Zangwill (eds.), *Current Problems in Animal Behaviour.* Cambridge: Cambridge University Press.

Harlow, H. F. 1969. "Love in Infant Monkeys." *Scientific American,* 200(6):68-74.

Harlow, H. F., and M. K. Harlow. 1962. "Social Deprivation in Monkeys." *Scientific American,* 207:137-146.

Hertz, R. 1909. *Death and the Right Hand* (the 1960 edition) Aberdeen: Cohen and West.

Holloway, R. L., Jr. 1966. "Cranial Capacity, Neural Reorganization, and Hominid Evolution: A Search for More Suitable Parameters." *American Anthropologist,* 68:103-121.

——1968a. "Cranial Capacity and the Evolution of the Human Brain." In M. F. A. Montagu (ed.), *Culture: Man's Adaptive Dimension.* New York: Oxford University Press.

——1968b. "The Evolution of the Primate Brain: Some Aspects of Quantitative Relations." *Brain Research,* 7(2):121-172.

Hubel, D. H., and T. N. Wiesel. 1962. "Receptive Fields, Binocular Interaction and Functional Architecture in the Cat's Visual Cortex." *Journal of Physiology.* 160:106-154.

John, E. R., R. N. Herrington, and S. Sutton. 1967. "Effects of Visual Form on the Evoked Response." *Science,* 1955: 1439-1442.

Kimura, D. 1973. "The Asymmetry of the Human Brain." *Scientific American,* 228(3):70-78.

Kleist, K. 1934. *Gehirnpathologie.* Leipzig: Barth.

Koenigswald, G. H. R. von. 1962. *The Evolution of Man.* Ann Arbor: University of Michigan Press.

Kotchetkova, V. I. 1970. "On Brain Size and Behavior in Early Man." *Current Anthropology,* 11:176.

Kraft, M. S., W. D. Obrist, and K. H. Pribram. 1960. "The Effect of Irritative Lesions of the Striate Cortex on Learning of Visual Discriminations in Monkeys." *Journal of Comparative Physiology* 51:17-22.

Lashley, K. S. 1947. "Structural Variation in the Nervous System in Relation to Behavior." *Psychological Review,* 54:325-341.

Lashley, K. S., and G. Clark. 1946. "The Cytoarchitecture of the Cerebral Cortex of Ateles: A Critical Examination of Cytoarchitectonic Studies." *Journal of Comparative Neurology* 85:223-306.

Le Gros Clark, W. E. 1963. *The Antecedents of Man.* New York: Harper and Row.

——1964. *The Fossil Evidence for Human Evolution.* Chicago: University of Chicago Press.

Lehman, R. A. W. 1970. "Hand Preference and Cerebral Predominance in 24 Rhesus Monkeys." *Journal of Neurological Science,* 10:185-192.

Levy, Jerre. 1972. "Lateral Specialization of the Human Brain: Behavioral Manifestations and Possible Evolutionary Basis." In *The Biology of Behavior* (ed. by J. A. Kiger). Corvallis: Oregon State University Press.

Levy, J., and C. D. Laughlin. 1974. "Neural Pathways for Manual Control and Method of Writing in Left and Right Handers." (In preparation.)

Levy, J., and J. Mandel. 1972. "Lateral Field Differences for Verbal and Non-Verbal Material in Sinistrals." n.p.

Levy, J., and T. Nagylaki. 1972. "A Model for the Genetics of Handedness." *Genetics,* 72:117-128.

Lex, B. 1973. "Biological Interpretations of Ritual Trance and Ritual Therapy: Rethinking Altered States of Consciousness." n.p.

Luria, A. R. 1966. *Higher Cortical Functions in Man.* New York: Basic Books.

Nebes, R. D. 1971. "Superiority of the Minor Hemisphere in Commissurotomized Man for the Perception of Part-Whole Relations." *Cortex,* 7:333-349.

——1972. "Dominance of the Minor Hemisphere in Commissurotomized Man on a Test of Figural Unification." *Brain,* 95:633-638.

Nebes, R. D., and R. W. Sperry. 1971. "Hemisphereic Deconnection Syndrome with Cerebral Birth Injury in the Dominant Arm Area. *Neuropsychologia,* 9:247-259.

Nottebohm, F. 1971. "Neural Lateralization of Vocal Control in a Passerine Bird I. Song." *Journal of Experimental Zoology,* 177:229-262.

——1972. "Neural Lateralization of Vocal Control in a Passerine Bird II. Subsong, Calls, and a Theory of Vocal Learning." *Journal of Experimental Zoology,* 179:35-50.

Ornstein, R. E. 1972. *The Psychology of Consciousness.* San Francisco: W. H. Freeman.

Penfield, W., and L. Roberts. 1959. *Speech and Brain-Mechanisms.* New York: Atheneum.

Pribram, K. H. 1971. *Languages of the Brain: Experimental Paradoxes and Principles in Neuropsychology.* Englewood Cliffs: Prentice-Hall.

Pribram, K. H., and A. R. Luria. 1973. *Psychophysiology of the Frontal Lobes.* New York: Academic Press.

Robinson, J. T. 1962. "The Origin and Adaptive Radiation of the Australopithecines." In G. Kurth (ed.), *Evolution and Hominization,* pp. 120-140.

Shankweiler, D., and M. Studdert-Kennedy. 1967. "Identification of Consonants and Vowels Presented to Left and Right Ears." *Quarterly Journal of Experimental Psychology,* 19:59-63.

Shellshear, J., and G. Elliot Smith. 1934. "A Comparative Study of the Endocranial Cast of *Sinanthropus.*" *Philosophical Transactions of the Royal Society of London, B.,* 223:469.

Sperry, R. W., N. Miner, and R. E. Meyers. 1955. "Visual Pattern Perception Following Subplial Slicing and Tantalum Wire Implantations in the Visual Cortex." *Journal Comparative Physiology,* 48:50-58.

Sperry, R. W., M. S., Gazzaniga, and J. E. Bogen. 1969. "Interhemispheric Relationships: the Neocortical Commissures; Syndromes of Hemisphere Disconnection." In *Handbook of Clinical Neurology* (P. J. Vinken and G. W. Bruyn, eds.), Vol. 4, pp. 273-290. Amsterdam: North Holland Publishing Co.

Stamm, J. S., and K. H. Pribram. 1961. "Effects of Epileptogenic Lesions of Inferotemporal Cortex on Learning and Retention in Monkeys. *Journal of Comparative and Physiological Psychology,* 54:614-618.

Stephan, H., and O. J. Andy. 1966. "Quantitative Comparisons of Brain Structures from Insectivores to Primates." *American Zoologist,* 4:59-74.

Trevarthen, C. 1962a. "Exploring the Neural Mechanisms of Mind." *Engineering and Science Magazine,* November.

——1962b. "Double Visual Learning in Split-Brain Monkeys." *Science,* 136:258-259.

——1969. "Brain Bisymmetry and the Role of the Corpus Callosum in Behavior and Conscious Experience." Presented at the International Colloquium on Interhemispheric Relations, Czechoslavakia, June 10-13.

Warren, J. M. 1953. "Handedness in the Rhesus Monkey." *Science,* 118:622.

Weigl, E. 1927. "Zur Psychologie der sogenannten Abstraktionsprocesse." *Zeitschrift für Psychologie,* 103.

Zucker, K. 1934. "An Analysis of Disturbed Function in Aphasia." *Brain,* 57.

IV

Cognitive Extension of Prehension

> Though it be too obvious to escape observation, that different ideas are connected together; I do not find that any philosopher has attempted to enumerate or class all the principles of association; a subject, however, that seems worthy of curiosity. To me, there appear to be only three principles of connection among ideas, namely, *Resemblance, Contiguity* in time or place, and *Cause* or *Effect*.—DAVID HUME, *An Enquiry Concerning Human Understanding*

ALFRED NORTH WHITEHEAD isolated two aspects or modes of prehension.[1] The first and most apparent to awareness he termed "presentational immediacy" (1957, p. 101). We will call this mode simply the mode of "simultaneity" in prehension. It is the aspect from which we principally derive our impression of concrete objects, classes of objects, and simultaneous relationships between entities in the real world. The relationship between a boy and a ball that is in his hand is an example of data derived in this mode. It may be seen

[1] Whitehead defines prehension to include three factors: "(a) the 'subject' which is prehending, namely, the actual entity in which that prehension is a concrete element; (b) the 'datum' which is prehended; (c) the 'subjective form' which is *how* that subject prehends that datum" (1957, p. 35). Following Whitehead, and assuming a *Homo sapiens* or other higher organism as subject, we may define prehension as the inputting of data through any of the sensory receptors into the brain plus the simultaneous association of those data with other data and associations of data stored in memory within the brain.

as the mode of spatial relations, common-sense direction, contiguity, and the like.

The second of Whitehead's prehensional modes, and the least apparent to awareness, is the mode of "causal efficacy" (1957, pp. 184 ff.). We will hereafter call this the mode of "causality." Through this mode we derive our sense of the flow or process of patterned change, of cause and effect, of the evolution of forms. The boy throws the ball, and a second later we hear the tinkle of breaking glass. We prehend, almost without reflection, the singleness of process, the cause and the effect. "The boy has broken a window." What is critical to understand here is that simultaneity and causality are but two aspects of a single process—*prehension*—and that this process is common, though crude or complex by degree, to all sentient beings.

That prehension is common to all sentient beings, and that it will likely be common to all artificial intelligences in the future, is best demonstrated by Michael Scriven in a discussion of what he terms the "comprehension theorem:"

> . . . we can see that two keys to handling (over-) rich input selectivity in allocating storage space and use of radically economical filing systems. The main trick consists in *creating* redundancies in the input. What can be counted as a redundancy obviously depends on the utility program/agent. What makes a datum redundant may be the trivial discovery of direct or approximate duplication of the datum-watching for this is what makes the ordinary redundancy check . . . useful. But this cannot cope with a rich information environment. To do that, we have to discover extrapolable regularities in the input. If there are none, the environment is totally chaotic and cannot be handled once capacity is reached (or before). But the slightest long-term deviation from randomness creates the possibility of using the (stochastic) pattern involved to render a large or infinite amount of input redundant since it can be generated internally from the stored pattern. So pattern-making and pattern-recognizing become the lifelines of the drowning IPD (information-acquisition-storage-retrieval-reaction device). The fundamental question for the evaluating routine is whether the mismatch between a new datum and what is already stored, or can be internally generated, is more serious than the use of resources involved in storing the exact datum . . . . in particular the comprehension theorum claims that comprehension is nothing more than a dispositional capacity side effect of an effective information storage and retrieval system in a com-

> plex, open-ended environment. An environment is open-ended if its content increases with time or closer examination; it is complex if its features require more than a *few, simple, predictable,* regularities to describe it completely. (n.d., pp. 35-36)

Scriven is specifically addressing himself to the probable attributes of supercomputers. However, he is perfectly aware that the theorem applies to all living systems as well. All sentient beings exist in an "over-rich data environment"—that is, no organism is equipped with memory capable of storing all sensory input. More importantly, all such beings are presented with environmentally based decisioning problems—i.e., they must be capable of selectively responding to input in an adaptively appropriate way (a point also made by Goodson, 1973, p. 29). Responding to each event as a unique occurrence would be clearly maladaptive, for random response to environment would quickly end the career of each and every member of a population. Hence all species must, for survival, be capable of discriminating between relevant and irrelevant input.

Species distinguish relevant from irrelevant patterns by constructing models of their environment against which they match sensory input for relevance, from which they generate expectations of the state of their environment, and that they modify as the sensory input requires. Models for most organisms are located in sensory association areas, specific exceptions being man and chimpanzee where models may span two or more association areas.

As noted in chapter 3, there is a large body of evidence indicating that such models are not composed of a small discrete number of cells, one cell group representing one model. John (1967) has brilliantly reviewed the evidence relating to this question and concludes that "information in the nervous system is conceived of, not as the deterministic discharge of certain coded cells which 'stand for' an item of information, but as the spatiotemporal patterns of organization in enormous aggregates of neurones" (p. 419).

The construction of models is ultimately based upon information about causal and spatial relations presented through sensory input and "built up" through ontogeny in time. As Levi-Strauss has put it, there are homologous relations between entities in the world and entities (elements) in structural models. Yet the memory of no organism is a blank slate at birth. At least some, if not many, of the neural associative structures are programmed genetically—are passed from generation to generation through genetic material. The phenomenon known as imprinting is one apparent example of this process. Another likely example is the universal stages in the acquisition of language in children (Lenneberg, 1967, chapter 7; Slobin, 1971).

Holographic Models

The most satisfactory theory to date that may help explain how the central nervous system forms models of reality has been amplified recently by Karl Pribram (1971a, 1971b), and is grounded upon an intricate analogy with holography:

> Holography is a kind of photography or in more sophisticated terminology, an optical information processing mechanism. But holography differs radically from normal photography. In a regular camera the film records the intensity of light that is reflected from objects. Each point on the film stores information from a single corresponding point in the photographed scene. The resulting picture looks like the original scene. In holography, light from every point in the scene is distributed (diffused) to many points in the film. When the film is developed, no visible pictures or images appear. Instead the film has a pattern of tiny and larger swirls, interference patterns that look like a piece of moire silk. (Pribram, 1971a, p. 88)

It is not the objective pattern, or "picture," of reality presented to the senses that is recorded in memory, but rather something on the order of an "interference pattern" that is recorded.[2]

The holographic hypothesis seems to solve a number of paradoxical findings in neurophysiological research, not the least of which is the fact that memories seem to be recorded over wide

[2] In holography the scene is illuminated by a laser that is a beam of coherent light. The image that reaches the plate, however, has been disturbed—the light is no longer coherent. Combined with the coherent light, the mixed beam has a characteristic pattern of interference. It is this pattern, rather than the scene, that is recorded (see Gabor et al., 1971).

expanses of the cerebral cortex and are often left unimpaired by massive destruction of brain tissue (Pribram, 1971b, pp. 116 ff). A fascinating characteristic of a holograph is that the entire interference pattern of the subject scene is recorded simultaneously on all parts of the plate. Thus the entire scene may be projected from a mere fragment of the plate (Gabor, 1971, p. 15). For another thing, the scene is recorded in three dimensions, not just two as with most photography, thus recording total parallax. Finally, the holographic process allows a large number of patterns to be recorded on the same plate. "Some 10 billion bits of information have been stored holographically in one cubic centimeter" (Gabor, 1971, p. 89). In holographic "photography" hundreds of individual scenes may be recorded on the same plate (Asimov, 1972b, p. 32). This factor would explain the enormous storage and retrieval capacity of our memory circuitry, a point similarly implied in a recent discussion of the brain by Isaac Asimov (1972a, p. 37).

An uncritical application of the holography analogy to an understanding of higher cortical modeling in man might lead to at least two misunderstandings. In the first place, we do not mean to imply that holography as a process of laser physics is equivalent to modeling in the brain. It is an interesting fact in the history of the neurological sciences that, at every stage, the brain has been likened to the most complex technology available at the time. During the last century the brain was compared with a complex clock. During the early decades of this century, it was seen as a kind of switchboard. With the development of computer technology, the temptation to compare the central nervous system and electronic "brains" became irresistible. Thus, when we resort to an analogy from the study of holography, we are continuing a time-honored, but limited, tradition. We must emphasize, of course, that neural models are not the same thing as holographs; they are not recorded in the same medium and are not the results of the same mechanisms. Furthermore, neural models are an infinitely more complex phenomenon than the holograph.

In the second place, an uncritical appeal to the holography analogy might lead one to take the position of the generalists (see chapter 3). You will remember that the generalists have held that memories are recorded at all points of the cortex and are never localized. We have shown that the evidence does not support this position and would argue that the "holographic" quality of memory may apply as well to the recording of memory within and between primary association areas. It seems likely that the holographic hypothesis is especially useful with reference to the identification function of the visual association areas, to the inferior parietal lobule, and to the association area (Brodmann's areas 18 and 19). The hypothesis may also apply to the other sensory association areas, to the inferior parietal lobule, and to the "engrams" of the motor cortex.

With these reservations aside, the holographic analogy helps to shed light to some extent on the nature of neural models. Probably the most compelling feature of the hologram is that it provides an example of recorded and retrievable information about sensory experience that is lodged in a physical "code"; and a code that is totally dissimilar to the nature of the experience itself. Pribram, (1971a, 1971b) is likely correct in hypothesizing that memory functions much like a holographic plate upon which may be stored millions of discrete models. Henceforth, we shall term all neural models simply *holographic models.*

In attempting to explain how the model of the physical hologram can be applied to information storage in the brain, one must first demonstrate the probability of the formation of the interference patterns necessary to generate a hologram. Pribram (1971b) positively eschews any theories of "disembodied floating fields" or even "disembodied wave forms." He quite reasonably looks to the microstructure of the slow postsynaptic potentials as the locus of the requisite interference patterns. He states:

> . . . consider once again the construction of more or less temporary organizations of cortical columns (or, in other neural locations, other aggregates of cell assemblies) by the arrival of impulses at neuronal junctions which activates horizontal cell inhibitory interactions. When such arrival patterns converge from at least two sources their designs would produce interference patterns. Assume that these interference patterns made up of classical postsynaptic potentials are coordinate with awareness. Assume also that the microstructure of slow potentials is accurately described by the equations that describe the holographic process which is also composed of interference patterns. The conclusion

> then follows that information representing the input is distributed over the entire extent of the neural pattern just as it is over the entire extent of the physical holographic pattern. There is, however, considerable leeway in how one can currently conceive of this distribution and correlation process. (pp. 152-153)

We would agree that this is the most probable mechanism to explain the generation of interference patterns essential to the holographic hypothesis. But we feel that there is more to the model-forming process than simply what is stated in the paragraph quoted above. That may be sufficient to explain simple perception. But setting up a relatively stable model of reality probably involves the actual physical growth of new neural connections, which in turn may generate more complex interference patterns following Pribram's general outline.

The storage of a model depends literally upon the physical growth of nerve cells; more specifically, a model is physically comprised of a configuration of dendritic-synaptic structure (Cf., Eccles, 1973: 176 ff). The recording of a model will generally involve neurons over a wide area of cortical tissue, and a change in a holographic model will inevitably entail the physical alteration of the structure. Thus there must always be a time lag between the input of sensory data necessitating an alteration in a holographic model and the completion of the alteration itself. Also, a model of data derived through two or more sensory modes will require neurophysiological connections between partial models lodged in the primary association areas. For instance, the model of a friend will include partial models corresponding to the sound of his voice, his visual image, and perhaps his particular odor. In man, at least, these separate submodels will usually be linked to form a single model that transcends any primary association area.

Though we have characterized the probable form of holographic models, the question still remains as to where they originate and reside in the human brain. The most fundamental models, as with imprinting, are structured genetically and thus provide a biogenetic base for the organization of sensory input. More specifically, the genetic basis of models has two aspects: (i) the most fundamental models are, at birth or at the appropriate stage of maturation, comprised of genetically derived "content" or *neurognosis,* and (ii) the structural mechanisms for ramifying the extent of neurognostic models are genetically determined. Of these two aspects, neurognosis requires some amplification. Neurognosis is, as the term implies, biogenetically given rudimentary information about the world, embedded in various association areas in the brain, either by the time of birth, or at some later maturational stage. The neurognostic model (a type of "holographic" model) may be viewed as something like the first approximation (or model "seed") of phenomena in the world, models later to be modified, amplified, and even circumvented through learning. The more simple and obvious examples of neurognostic modeling are the neurophysiological "latent processes" underlying the behaviors of sucking and grasping in the newborn, and of language acquisition and perhaps homosexuality at later stages of maturation. Furthermore, we contend that the neurognostic models form the initial "crystals" around which form many other behavior patterns that, like music, primary social group identification, and conception of repetitive time, appear to be universal, or virtually universal, among human societies.

We are not saying that neurognostic models are precisely as they were at various prehominid stages. Like any other aspect of the phenotype, they will reflect genetic variation and selection in the past in favor of adaptive formations. Other models are, of course, derived solely through the synthesis or inductive organization of sensory input (*à la* the Scriven theorem). Some, if not most, of the models with which we are here concerned are those the cores of which were genetically given, only later to be modified by sensory input and association. It requires emphasis that neurognostic models are not immutable but are usually modified to fit the peculiar experience of the particular organism (see Rosenzweig et al., 1972).

To put this another way, for neurognosis to come to fruition, at whatever stage of maturation it is initiated, requires a sequence of ontogenetic "feedback" from reality. There is increasing evidence for man and animals alike that deprivation in ontogeny will inhibit the sequential development of neurognosis. And this effect may be felt on neurognosis at all levels of complexity in the central nervous system, from the simple reflex to the higher cortical functions (see the language acquisition section in chapter 7, and the section on "critical periods" in chapter 8). As Jean Piaget (1971) has said:

> At every stage of development of the living structure it is clear that, in fact, the essential question is that of the regulation mechanisms. At the physiological level of the synergy of functions, the problem underlying all these questions is that of homeostasis: the equilibrium of each open system and the hormonal or nervous regulations which assure coordination in the whole living entity. At the stage of ontogenetic development . . . the central problem is that of the dynamic equilibrium of *"channeled" formations—"homeorhesis," which Waddington has rightly differentiated from homeostasis.* (p. 12)

"Homeorhesis," for Piaget, is homeostasis with duration. That is, the structure (neurognosis) is in a state of "equilibration" with external reality, and "autoregulation" between structures (Piaget, 1971, pp. 10 ff). It is with the question of equilibration that we are here concerned, and for a solution to this problem we now turn to a discussion of the empirical modification cycle.

Empirical Modification Cycle (EMC)

How does the organism, *Homo sapiens* or any other species with a central nervous system, modify its models? Virtually all organisms exhibit a process that may be termed a *cycle of empirical modification* (EMC). That is, the holographic models (usually combining inherited and acquired components) are continually being matched against patterns of sensory input. The brain carries out this function by generating *probability expectations* upon which the behavior of the organism *vis à vis* its environment will be based. In this fashion, the organism "tests" the accuracy of its holographic representations of reality, generally through the medium of behavior or motor response, against feedback from reality. Input data are matched against the probability expectations. If the "fit" between input and expectation is within adaptive limits, the input data are released from memory. If, on the other hand, the input data and expectations do not match, the data are retained and utilized in the modification of the underlying holographic model (see discussion of Sokolov's work on habituation in Pribram, 1971a, p. 48).[3]

The behaviorial condition we term "wakefulness" corresponds closely with optimal activity in the empirical modification cycle. What occurs during sleep will be discussed in a later chapter. As an aside, researches into the nature of "subjective probability" in studies of human decisioning may in the end be concerned with the probability expectation aspect of the empirical modification cycle (see Edwards and Tversky, 1967, Part 2).

We wish to stress again that the process of empirical modification is one of physical change in the dendritic-axonic-synaptic structure of long-term memory circuitry. Experiments with the effects of controlled environments on the physical-chemical structure of the brains of rats offer indirect support for the neurophysiological structure of models and the type of modification of these advocated in the present theory. Mark R. Rosenzweig, Edward L. Bennett, and Marian Cleeves Diamond (1972) subjected three groups of rats to three types of environment—the standard laboratory cage, an "enriched" environment with more than normal novel input, and an "impoverished" environment much like solitary confinement (p. 22). A number of discoveries were made when the brains of the rats were examined after they had lived in these environments for some time:

> . . . we had found that rats that had spent from four to ten weeks in the enriched or the impoverished environments differed in the following ways: rats with enriched experience had a greater weight of cerebral cortex, a greater total activity of acetylcholesterase but less activity of the enzyme per unit of tissue weight . . . there were more glial cells

[3] Our conception of the EMC is similar in some respects to the notion of TOTE (test-operate-test-exit process) that has been independently formulated by Pribram (1971b, pp. 90 ff). Pribram defines TOTE as ". . . a logic 'element' made up of two reciprocally related processes: (1) the 'test,' a state composed of junctional (synaptic and dendritic) potentials, and (2) an 'operation' on that state by nerve impulses generated by receptors or by the central nervous system" (p. 95). The two conceptions emphasize the operation by which input and response are compared for appropriate fit. The two conceptions differ in that the EMC emphasizes the expectation or prediction operation. The reader is referred to chapter 6 for a more detailed explication of the importance of the predictive operation in the modification of models.

> in rats from the enriched environment than there were in rats from the impoverished one. . . . Although differences in experience did not change the number of nerve cells per unit of tissue, the enriched environment produced larger cell bodies and nuclei.
>
> . . . The most consistent effect of experience on the brain that we found was the ratio of the weight of the cortex to the weight of the rest of the brain: the sub-cortex. It appears that the cortex increases in weight quite readily in response to an enriched environment, whereas the weight of the rest of the brain changes little. (Rosenzweiz et al., pp. 24 and 25)

These scientists also report a number of other studies that indicate a physiological base for memory in rats:

> . . . there is increased synaptic contact in enriched-experience rats . . . increased size of synapse is associated with a decreased number of synapses, whereas decreased size of synapse is associated with an increased number. It seems that memory or learning may be encoded in the brain either by the selective addition of contacts between nerve cells or by the selective removal of contacts, and that both processes may go on at the same time. (pp. 27 and 28)

Traditional Learning Theory

It is appropriate both for explicating the process by which models are acquired and for our later criticism of culture theory to reexamine the concept of "learning" from the biogenetic structuralist point of view. Orthodox learning theory recognizes the existence of two types of learning:

(i) One fundamental type of learning is termed *classical conditioning* and occurs when an unconditioned stimulus (like food) is paired with a conditioned stimulus (like a light or a bell) in the presence of a natural, or "built-in" response (like salivation). After a while, the unconditioned stimulus is removed, leaving only the conditioned stimulus, which continues to elicit the response. Learning in classical conditioning involves a natural response that is usually mediated via the autonomic nervous system. The "learning" consists of switching the stimulus from a natural to an arbitrary, unnatural one by the simple device of repeated pairings.

(ii) A more sophisticated type of learning is termed *instrumental* or *operant conditioning* and involves the selective reinforcement (if "positive," then pleasurable or rewarding; if "negative," then removal of unpleasurable or aversive stimulus) of one of a set of alternative responses to a conditioned stimulus. To give an example, a bell may be rung (the conditioned stimulus). When and if the subject selects the desired one of a set of (say) three responses, he is given food as a reward (reinforcement). If this procedure is repeated a number of times, the subject will continue for a number of trials to repeat the desired response to the conditioned stimulus, even in the absence of the initial reward. The subject is said to be instrumentally conditioned.

In other words, the subject has learned the "correct" response, i.e., the response desired and reinforced by the experimenter. Eventually, of course, the response will extinguish if all reinforcement (food in this case) is stopped permanently. In instrumental (operant) conditioning the response (e.g., pressing the "correct" bar, pushing the "correct" button) is mediated via the so-called voluntary nervous system rather than via the autonomic nervous system as in classical conditioning (response-salivation in the above example). Another difference is that in classical conditioning the reinforcer (here food) *precedes* the conditioned response (salivation). In instrumental conditioning, the reinforcer (here also food) *follows* the conditioned response (pressing the correct bar.)

As the reader may already be aware, "learning" as defined by orthodox learning theorists involves input (stimulus) and output (response) with virtually no reference to structuring and associative processes internal to the brain. The brain, in point of fact, has often been treated as a "black box," an unknown quantum. Hence it is usually ignored in the learning formulation. Only in the last few years have learning theorists even talked about "private events," that is, subjective events or processes like thought, feeling states, and decisioning. Yet we suspect there are few, if any, learning psychologists who would deny that learning involves neurophysiological processes and structuring at some level. And, as will be mentioned in relation to prepared learning and psychopathology (chapters 5 and 8), some theorists are willing to concede at least gross effects of brain structures on certain types of learning situations and resultant responses.

To continue with our discussion, when we refer to a "positive reinforcer" we are indicating a medium for the production of pleasurable sensations similar to those elicited by direct stimulation of the median forebrain bundle, part of the reticular-limbic system. This area is often called the "pleasure center" and is the site of the strongest reinforcement owing to its producing a state of marked pleasure or contentedness in the subject. It is quite true that *classical* conditioning, depending as it does upon autonomic connections, does not directly rely, under normal experimental conditions, upon reinforcement through the limbic system. Yet, as Neal E. Miller of Rockefeller University and, more recently, Leo V. DiCara (1970) have shown, even the autonomic system may be, under special experimental conditions, instrumentally conditioned. Their contention is that the two types of learning are, in fact, two manifestations of the same type of learning. It is not our intention to take sides in this controversy, but rather to indicate that whether one type or two, learning of a "permanent" or long-term nature involves some modification of the physical structure of the neuron net comprising holographic models. This may be through change in biochemistry, through alteration of membrane potential, variation at the synapse, change in the additivity function at the initial segment, increase or decrease in dendritic-axonic branching, and the like.

From the biogenetic structuralist's point of view, *learning* is any activity along the *empirical modification cycle* that results in at least temporary formation or reformation of a "holographic" model. Thus new input either confirms a preexisting model or else results in the modification of a preexisting model. If modified, whether the modification is major or minor depends on the degree of "fit" of the new input. In any case, biogenetic structuralism predicts that the models, however modified, are retained even after their behavioral manifestations are no longer overtly manifested and reinforced. The exception to this statement of permanence would be the brief encoding of new information in what has been called "short-term" memory, which probably involves molecular changes rather than structural modification of neurones. The content of short-term memory is either incorporated into one of the models of reality we have been discussing or else it is apparently lost. In other words, acquired models may be relegated to the level of acquired *latent* processes after reinforcement of the behavior has been removed. It is a not uncommon phenomenon in experimental research for a response (especially if aversive), which has once been extinguished, to reappear spontaneously under certain conditions, although the initial aversive stimulus remains inoperative. Even if a response does not spontaneously reappear, it takes far fewer trials, using the initial stimulus to reestablish the response after extinction (see Osgood, 1953, pp. 337 ff.).

As we have seen from previous discussions of associative processes in general, learning is clearly a more complex phenomenon in man than traditional learning theory would lead one to suppose. Man and, to some extent, the chimpanzee are capable of cross-modal transfer. That means he is able to associate information derived through two or more sense modalities, generating in effect composite models of reality independent of any particular sense modality. Furthermore, learning in man may take place without immediate mediation of the limbic-reticular system. He is able to acquire and modify models in the absence of immediate external positive or negative reinforcement. This is not to say, of course, that man does not learn by instrumental means, but rather that this may take the form of auto-reinforcement (self-excitation of the limbic system) as well as the form of the traditional external reinforcement. Furthermore, human learning can avoid immediate limbic reinforcement via cross-modal transfer of information.

We have implied above that memory is divided into surface and deep levels. Actually, we do not wish to imply that all "deep" memory models in the brain have the same stability. As a matter of fact, the reverse is most likely the case. The assertion of two levels of memory may be a shorthand fiction. The stability of models probably depends greatly upon the area in which they are embedded. There is evidence, for example, that complex models located in the neocortex are more tenuous. In fact, this area as a whole is more sensitive than other areas to noxious stimuli. The effect of barbiturates, general anesthetics, carbon monoxide, or anoxia when introduced to the central nervous system is a sequential decriment of neural function beginning with the neocortex and extending down to the level of the thalamus and brain stem and finally to the level of the spinal cord, the latter region being relatively resistant to such effects.

There remain many intriguing questions pertaining to the nature and function of neural models and structures, certainly more questions at this point than answers. One of the more interesting questions centers on the neurophysiological basis of "loss of memory." When we fail to retrieve a bit of previously modeled information, does this mean the model has been in some fashion disassembled, or has it been "stored" in a memory bank for which there is limited access? If the latter, what are the laws governing access to the memory bank (or banks)? How is a student confronting an examination able to recall a bit of information he "learned" the night before and be incapable of recalling that very same bit of information a week or two later? And, finally, how is that very same student able to recall, seemingly at will, a trivial incident occurring ten or more years previously?

Cognitive Extension of Prehension (CEP)

To return again to the evolution of the human brain: as already indicated, the long and gradual increase in the size and dendritic-axonic-synaptic complexity of this organ reached a critical point at some time prior to the australopithecine evolutionary phase. At this time there occurred one of the planet's most interesting and important natural revolutions. We may call this revolution the *cognitive extension of prehension* or, abbreviated, CEP.

Prior to the occurrence of CEP, the prehominid brain was incapable of prehending, either in the mode of simultaneity or the mode of causality, much beyond what may be called *common sense.* Common-sense relations may be defined as those between events and objects in either mode that are completely present in the sensory field (see Figure 4.1). Contiguity of relationship in time and

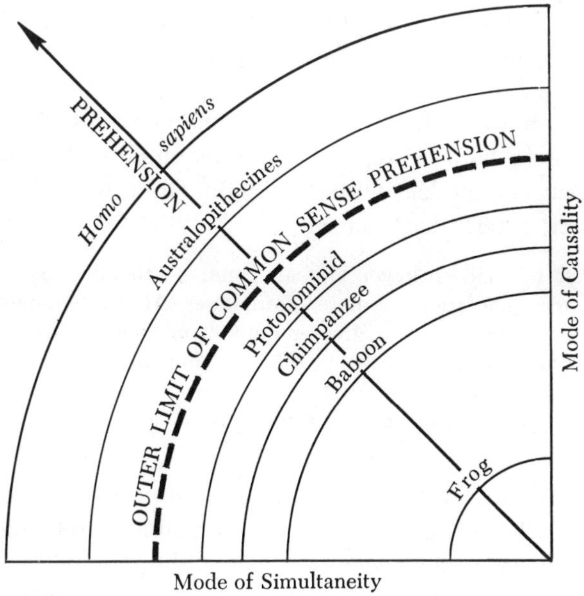

Figure 4.1. The Cognitive Extension of Prehension

space, combined with the lack of spontaneous cross-modal transfer, defined the outer bounds both of prehension and of neurognosis within the brain of our prerevolutionary ancestor.

Finally, in response to the allometric pressure of increased body size, the size and complexity of the neurons reached an accretive, but qualitative, shift in function. The hominid brain became capable of prehending *beyond* the bounds of common sense, associating both spatial relationships that were not immediately present in the field of prehension and causal sequences involving events not immediately contiguous in time. Though the neural capacity must have been minimal at first, the shift in capacity of the temporal or causal aspect was to prove the most crucial because of the potent selective advantage of the predictive function:

> One can easily see how such an evolutionary development had extreme survival value. By registering the elements of the environment into conceptual and causal forms, primitive man could evaluate dangers not immediately present, and he could form some expectations as to the results of his actions. (d'Aquili, 1972, p. 25)

What occurred at this time amounted to an extension of the bounds of cognition beyond relationships immediately prehended in the sensory field to relationships more expansive in time and space. This extension was reflected in the increased complexity of holographic models, both neurognostic and acquired, and associational relations between models within the brain. Or, as Schroder, Driver, and Streufert (1967) have put it:

> Compared to that of lower animals, human "thought" is characterized by the generation of more alternatives. More meanings can be attributed to objects and a greater number of connections (relations) between these meanings arise. In this way, human thought is less stimulus bound; action can be delayed; a given stimulus gives rise to a greater number of outcomes, creating more uncertainty and ambiguity. (p. 5)

"More uncertainty and ambiguity" is an introspective way of saying that because the hominid brain became capable of relating objects and events in ever more complex models, the probability expectations generated from these models tended to be proportionally less and less representative of detailed reality. Feedback to the models became less immediate. Yet the extension of model complexity utilized the same associative tracts (plus some new ones) as before. They were subjected as before to the same empirical modification cycle and were modified utilizing the same dendritic-axomic material.

Lest we be charged with positing a "Rubicon" model for the evolution of human cognition and sociality, let us mention that the change we have termed the CEP was extremely complex, both behaviorally and neurophysiologically, and must have occurred gradually over a long period of time. But like many phenotypic changes corresponding to Cope's Rule, the gradual increment of change ultimately resulted in qualitative shifts in function of neuroanatomical subsystems, a point similarly made by Holloway (1968) for differential subsystem relations in different primate species.

It is now possible to indicate the ramifications of the CEP for the evolution of human society. Among other scholars, Washburn (1960, p. 49) has maintained that the marked increase in the size of the hominid brain was the effect of precultural man's acceptance of the tool as a weapon of defense, rather than his previous dependence upon large canines. Though an ingenious theory a decade ago, when little evidence of tool use by other species was known, this explanation of the origin of human techno-economics must now be viewed as no longer tenable. In the first place there is no evidence in the fossil record that man's prehominid male ancestors ever had inordinately large canines. In the second place the notion that acceptance of tool use was the prime cause cannot explain why man, alone among all the species we now know to utilize tools (see Van Lawick-Goodall, 1971, pp. 277 ff.), retained them and made them an integral part of his adaptive capability. A clearer view would be to see the retention of tools as an effect, rather than a cause. There exists solid evidence that "curiosity," instrumentality, and tool manipulation are linked to adaptation among forms having progressively complex brains. It seems to us, for example, that Menzel (1963, 1966) has shown that a greater tendency toward this behavior exists among chimpanzees than exists among macaque monkeys (*M. fuscata*). Protohominid cognition reached a point after the CEP when the creature was capable of predicting the utility of a retained hand-held bone, stone, or stick beyond the period of immediate use. Tool use and, more importantly, tool retention were the effects of selection for increased organizational complexity of the brain.

A similar theory for the linguistic origin of culture was advanced by Charles Hockett (1960), who argued that hominids developed a "protolanguage" consisting of concatenations of previously instinctive calls. The calls gradually lost their inherent meaning and became arbitrary, but meaningful, sounds (phonemes) from which true human language emerged. There followed as a consequence an increase in the size of the human brain. The same objections prevail against Hockett's theory as against Washburn's. In a word, complex vocal communication was the effect of selection for increased neural complexity. Of course, once established in the protohominid behavioral repertoire, both tool retention and complex vocal communication became distinctly and adaptively advantageous.

The precise sequence of evolutionary change that resulted in the development of language is uncertain. As will be suggested in a later chapter, it is possible that early hominids came to rely heavily upon vocal communication between infant and mother. As we have shown, there is clear indication that the development of language required the neurophysiological construction of tracts between primary association areas. If so, the capacity for spoken language required: (i) a link between Wernicke's area and Broca's area, these being associated with the emergence of the intersensory integration function of the dominant inferior parietal lobule, and the discrimination functions of the frontal processes, and (ii) the capacity for cross-modal transfer that permits the construction of classes of objects where the information formerly available was insufficient. It is true that brain lesions affecting speech areas (e.g., severance of the arcuate fasciculus) in the early years of maturation may be circumvented by the development of new neural tracts (see the results of Landau et al. in Lenneberg, 1966, p. 236). But most assuredly, no organism lacking the initial tracts would be capable of forming the secondary tracts and hence acquiring the faculty of speech.

Although the ramifications of biogenetic structuralism to the understanding of language acquisition will be examined in chapter 7, it is our feeling that the absolute distinction made between human capacity for language and the communication systems of other species is seriously misleading and requires some treatment here. One of the great contributions of Hockett's analysis, as well as subsequent analyses inspired by Hockett's work (see Altmann, 1967), has been the elucidation of the true complexity of human and non-human communication. This complexity is often described in terms of "design features," some of which have been described as unique to human language. Hockett (1960, pp. 8 ff.) pointed out that of the thirteen distinctive features characteristic of human speech, twelve are shared with other species. Only one—"duality of patterning," or the capacity to link meaningless sounds into meaningful utterances—he found unique to human speech. However, the uniqueness of duality of patterning to human speech has been challenged (Altmann, 1967, p. 347 ff.) and recent evidence collected from species of song birds suggests the use of this feature in the design of non-human vocal communication (see discussion of Smith's work in chapter 7). The point to be made here is that for a long time behavioral scientists have been in the habit of linking speech with human intelligence and have bolstered their bias by seeking design features that set human communication apart from communication utilized by the rest of the animal kingdom. Some authors have made reference to the fact that our closest relatives, the apes, cannot learn to speak. But, as we have seen, researchers have successfully taught language to chimpanzees utilizing modes of signaling other than vocalization (Gardner and Gardner, 1969; Premack, 1971a, 1971b; also see Brown, 1973, pp. 32 ff., for a relevant discussion of this issue). An overly zealous concern for design features may often, we feel, lead to attention to the surface, rather than the deep or universal, structure of communication. It may lead to various anthropocentrisms, all of which inhibit examination of factors, including neurophysiological infrastructures, upon an understanding of which truely explanatory models of communication must rest.

It seems likely that along with selection for a more complex brain as an allometric by-product of selection for bigger bodies, there also occurred a gradual increase in the selective coefficient for "intelligence." Intelligence is specifically defined here as the neurophysiological capacity for model reformulation and for novel (entrepreneurial) decisioning. We mean by this definition what Schroder el al (1967, p. 6) seem to mean when they state:

> Emergent rule structures . . . are exemplified by exploratory and creative behavior in animals at the upper end of the evolutionary scale, as well as by other forms of integratively complex thought in which many perspectives and ways of interrelating these perspectives occur, and in which *new* rules can be generated for decision-making purposes. Within this class also, systems differ in terms of the amount and speed of information processed, but they are similar in that new information-processing rules emerge within the system itself.

Selection in favor of intelligence would have accelerated in geometrical proportions once the capacity for entrepreneurial judgments (probably linked to evolution of the frontal lobes and to emergence of cross-modal transfer) became coupled with prehension beyond the bounds of common sense (i.e., the CEP). The evolutionary leap occurred when transitional man became neurophysiologically capable of the prediction that to retain a tool as a defensive weapon at time t_1 would give him an advantage when confronted by a predator, competitor, or prey at time t_2. The protolinguistic revolution occurred as a consequence of the capability for concepts or "classes" embedded in the neurognostic models that associated (i) input data from different sensory associative areas of the cerebral cortex and (ii) relations in time and space beyond the bounds of common sense. The problem of communication for the transitional man became the transferral of information about predictions not immediately apparent in the field of prehension, as well as coping innovatively with the predicted expectations. The result of hominid success at solving these problems is reflected in the increasingly complex patterns of adaptation to marginal environments apparent in the fossil record.

A major point to underline, and perhaps the major contribution this theory offers to the study of human evolution, is the orthogenetic continuity between the brain processes of prehominids and those of modern *Homo sapiens*. Though associative circuitry and the holographic models of reality embedded in neural structure became more complex in time, they did not, at any point, become independent of genotypic patterning. While man was increasingly able to predict events beyond the bounds of common sense, he did so in the same empirical way and utilizing essentially the same neurophysiological equipment as previously utilized. One of the most difficult aspects of the CEP to explain is how the holographic models characteristic of *Homo sapiens* came into being. The answer is that models inherited orthogenetically in prehominids were not simply erased with the advent of a "cultural revolution" as is often implied in cultural anthropological accounts. Far from it. They remain with man today, but in highly elaborated form. And where they once determined behavior in a more direct and impelling way, they now manifest themselves behaviorally as tendencies for particular sensory input to be associated in particular ways.

The latter point is critical. Primordial models inherent in the brains of protohominids were being tested by feedback through the empirical modification cycle. As indicated, adaptively significant models were kept adaptive by slight modification in response to negative feedback. Massively novel input either radically modified the model through positive feedback, or the particular configuration (hence, the organism) was selected against. The evolutionary trend was ever in the direction of maintaining adaptive change in the models. With the cognitive extension of prehension, the primary adaptive process did not change. Models continued to be constructed in the same fashion, and the negative and positive feedback aspects of prehension continued to function. What did change was (i) the expanse of reality controlled by models and (ii) the emergence of the phenomenon we often call "awareness."

The first of these shifts has already been discussed at length, and it is only necessary to mention a particular ramification of model expansion. As the duration of process and breadth of space controllable in models expanded, associations began to be made that had less and less adaptive importance. This accelerated as an unintended consequence of selection for adaptive intelligence and led to behavior patterns that were adaptively neutral. *Much that we view as social behavior, institutions, cultural values, etc., among contemporary human societies stem from models having for the most part no adaptive importance to the species or individual reproductive population.* Another way of saying this is that the CEP led to an optimal adaptive capacity, but the process of model building and elaboration continues and results in adaptively superfluous behavioral patterns, many of which we have learned to call "culture." Levi-Strauss is one of the few anthropologists to appreciate this point in his distinction between "elementary" and "complex" structures (1969, p. xxxiv). Complex social structures perform the same adaptive function as elementary social structures, i.e., they ally otherwise disparate social groups, but they combine this with other functions, some of which have little or no relation to the alliance function.

The second qualitative shift—the emergence of "awareness" and the sense of "self"—is an issue fraught with numerous philosophical and religious, not to mention scientific, complications (see Eccles, 1970). We will self-consciously restrict our discussion, therefore, to a brief comment relating the issue to the present consideration. The emergence of the model of "self" probably required a more sophisticated differentiation between the feedback originating from the proprioceptors and that coming from the exteroceptors (see chapter 5).[4] Furthermore, as Henry A. Murray has repeatedly emphasized, an individual's model of self is never static but requires a prehension of one's being as a temporal process. Our point should be clear at this juncture: the sense of "self" characteristic of contemporary human cognition is the direct result of the capacity to relate causal series extending both back far beyond the bounds of common sense, and forward in time through predictions (expectations, goals) that influence behavior in the present. Awareness, then, is the extremely heightened experience of direction, causality, and purposiveness of events internal and external to the organism.

References

Altmann, S.A. 1967. "The Structure of Primate Social Communication." In *Social Communication among Primates* (ed. by S. A. Altmann). Chicago: University of Chicago Press.

d'Aquili, Eugene G. 1972. "The Biopsychological Determinants of Culture." Reading: Addison-Wesley Modular Publications.

Asimov, Isaac. 1972a. "What's on Your Mind?" *Penthouse,* May, pp. 36 ff.

——1972b. "I Can't Believe I Saw the Whole Thing." *Saturday Review of Science,* September 2, pp. 25-32.

Brown, R. 1973. *A First Language.* Cambridge: Harvard University Press.

DiCara, Leo V. 1970. "Learning in the Autonomic Nervous System." *Scientific American,* 22(1):30.

Eccles, J.C. 1970. *Facing Reality.* New York: Springer.

——1973. *The Understanding of the Brain.* New York: McGraw-Hill.

Edwards, Ward, and Amos Tversky. 1967. *Decision Making.* Baltimore: Penguin Books.

Gabor, Dennis, et al. 1971. "Holography." *Science,* 173:11-23.

Gardner, R.A., and B.T. Gardner. 1969. "Teaching Sign Language to a Chimpanzee." *Science,* 165:664-672.

Goodson, F.E. 1973. *The Evolutionary Foundations of Psychology.* New York: Holt, Rinehart and Winston.

Hockett, Charles D. 1960. "The Origin of Speech." *Scientific American,* September.

[4] The proprioceptors are the sensors internal to the organism, whereas the exteroceptors are sensors giving the organism feedback from beyond its own boundaries.

Holloway, R.L., Jr. 1968. "The Evolution of the Primate Brain: Some Aspects of Quantitative Relations." *Brain Research,* 7(2):121-172.

John, E.R. 1967. *Mechanisms of Memory.* New York: Academic.

Lenneberg, Eric H. 1966. "The Natural History of Language." In *The Genesis of Language* (ed. by F. Smith and G.A. Miller). Cambridge: MIT Press.

——1967. *Biological Foundations of Language.* New York: John Wiley.

Levi-Strauss, Claude. 1969. *The Elementary Structures of Kinship.* Boston: Beacon Press.

Luria, A.R. 1966. *Higher Cortical Functions in Man.* New York: Basic Books.

Menzel, Emil W. 1966. "Responsiveness to Objects in Free-Ranging Japanese Monkeys." *Behavior,* 26:130-150.

Menzel, Emil W., et al. 1963. "The Effects of Environmental Restriction upon the Chimpanzee's Responsiveness to Objects." *Journal of Comparative and Physiological Psychology,* 56(1):78-85.

Osgood, C.E. 1953. *Method and Theory in Experimental Psychology.* New York: Oxford.

Piaget, J. 1971. *Biology and Knowledge.* Chicago: University of Chicago Press.

Premack, David. 1971a. "Some General Characteristics of a Method for Teaching Language to Organisms That Do Not Ordinarily Acquire It." In *Cognitive Processes of Nonhuman Primates* (ed. by L.E. Jarrard). New York: Academic Press.

——1971b. "Language in Chimpanzee?" *Science,* 172:808-822.

Pribram, Karl H. 1971a. "The Brain." *Psychology Today,* 5(4):44 ff.

——1971b. *Languages of the Brain.* Englewood Cliffs: Prentice-Hall.

Rosenweig, M.R., E.L. Bennett, and M.C. Diamond. 1972. "Brain Changes in Response to Experience." *Scientific American* (March), pp. 22-29.

Schroder, H.M., M.J. Driver, and S. Streufert. 1967. *Human Information Processing.* New York: Holt, Rinehart and Winston.

Scriven, Michael. n.d. "The Concept of Comprehension: From Semantics to Software."

Slobin, Dan I. 1971. "Developmental Psycholinguistics." In *A Survey of Linguistic Science* (ed. by W.O. Dingwall). Published by the Linguistics Program, University of Maryland.

Van Lawick-Goodall, J. 1971. *In the Shadow of Man.* Boston: Houghton Mifflin.

Washburn, S.L., and F. Clark Howell. 1960. "Human Evolution and Culture." In *Evolution After Darwin,* Vol. 11, *The Evolution of Man* (ed. by Sol Tax). Chicago: University of Chicago Press.

Whitehead, Alfred North. 1957. (1929) *Process and Reality.* New York: Harper Torchbooks.

V

Neurognostic Models: Structure and Content

> If pure thought runs all our trains, why should she run some so fast and some so slow, some through dull flats and some through gorgeous scenery, some to mountain-heights and jewelled mines, others through dismal swamps and darkness?—and run some off the track altogether, and into the wilderness of lunacy? Why do we spend years straining after a certain scientific or practical problem, but all in vain—thought refusing to evolve the solution we desire? And why, some day, walking in the street with our attention miles away from the quest, does the answer saunter into our minds as carelessly as if it had never been called for—suggested, possibly, by the flowers on the bonnet of the lady in front of us, or possibly by nothing that we can discover? If reason can give us relief then, why did she not do so earlier?
> —WILLIAM JAMES, *The Principles of Psychology*

IN CHAPTER 4 we considered the possibility that the brains of normally functioning adults possess a number of models of reality that are genetically preconditioned and against which sensory input is almost constantly compared. Quite another way of saying this is that sensory input is processed in different ways, depending upon the degree to which the input "fits" preexisting models. Thus sensory input may be said to be processed in a hierarchy of complexity. Be that as it may, we contend that models of reality and the channels through which they are judged for "fit" with the world are all comprised of real neuroanatomical material, and *only* such material.

Structure and Content

At this point, we should restate the conceptual distinction between *neurognosis,* or the informational content of models, and *neurognostic structure,* or the neurophysiological medium in which neurognosis is coded and through which it may be modified. Although the terms may elsewhere be used interchangeably, it is well for our present discussion to keep them distinct.[1] A neurognosis (neurognostic model) is a specific initial form in which reality is experienced. It is a model, as it were, that is minimally isomorphic with reality. It may be cognitive or affective, or both. It is often capable of being linked to (or "described by") some verbal utterance that itself may be simple or very complex. The verbal utterance, in turn, may have the attributes of a model of the neurognostic model. An example of one such simple formulation is "All events in the real world are ordered in causal sequences." This utterance is obviously not the neurognostic model itself, but it does form something of a shorthand description of the model, or a class of models.

A neurognostic structure, on the other hand, is the inferred neuroanatomical base for a neurognostic model. It need not be confined to one small area of the brain. Indeed, most neurognostic structures probably involve extremely complex synaptic relationships over wide cerebral areas, especially in *Homo sapiens.* It is perhaps easiest to think of neurognostic structure existing as a field of neuroanatomical pathways along which action potentials[2] proliferate in profusion, in marked contrast to a matrix of relatively latent, but similar, pathways (see Eccles, 1973, chapter 6). A neurognosis would be activated by a pattern of "firing" in a three-dimensional

[1] The distinction we are making here is directly analagous to the structure-function distinction fundamental to other areas of physiology.

[2] The term "action potential" refers to the electrical discharge propagated along the axon of a nerve cell. Action potentials are an "all-or-none" phenomenon. That is, once a neuron is excited sufficiently, it will fire with a fixed amplitude discharge which is unaffected by an increase in stimulation. If the excitation is below a critical, or "threshold," level, the neuron will not fire at all. The axon of a single nerve cell may connect with as many as thousands of other nerve cells, and an action potential propagated along that axon may participate in stimulating many of these other cells to fire, thus causing over time a vast field of action potentials.

array, more accurately, four-dimensional as any neurognostic activation would have duration. It is clear that each discrete neurognostic model has one, and only one, underlying neurognostic structure. Conversely, any discrete neurognostic structure presumably governs one, and only one, model. Yet it is equally clear that multiple models and structures may have, and almost certainly do have, overlapping pathways. That is, what distinguishes one model and structure from another is the total field of neural connections, not the particular cells, cell bundles, pathways, etc., that go to make up the field.

Neurognostic structures are so anatomically complex as to store models (in interference-like patterns) over large, though specific, areas of the brain. Thus a purely visual model of reality would be stored throughout most of the visual association cortex. This explains why large areas of the visual association cortex must be extirpated to produce a visual agnosia in animals. Anything less, and the animal will show minimal, if any, evidence of lack of visual recognition. In animals lacking the capability of cross-modal transfer, the presumption is that separate sensory models of reality are stored holographically in the association area of each sense modality. In man, and probably to some degree in the chimpanzee, integrated sensory models are at least partially stored in the dominant inferior parietal lobule. These models, by virtue of the integration of the sense modalities described in chapter 2, are "conceptual" in nature. It would be simplistic to presume, however, that models of reality exist passively in either the sensory association areas or the inferior parietal lobule. Insofar as the models contain dynamic relationships (either spatial or temporal), we can presume that the anterior convexity of the frontal lobes plays an important part in establishing and maintaining them. Hence the reciprocal connections between the anterior frontal area and the various sensory association areas and between the anterior frontal area and the inferior parietal lobule seem essential to the maintenance of the models of reality.

That relatively stable cognitive structures should exist on the sensory side of the central nervous system should not be surprising. It has been suggested for many years that motor analogues of these "sensory" structures exist. For example, any learned skilled activity, such as writing or tying one's shoes, is laid down in "motor engrams." An engram is defined as the definite and permanent trace left by repeated stimuli in neural tissue. A motor engram is generally understood to mean the stable neural basis or master plan for any learned skilled activity. Once an engram is activated, the skilled motor behavior is produced almost unconsciously. Some motor engrams, however, seem to be present at birth and are activated, partially at least, by progressive maturation of the central nervous system. Thus, although learning may be somewhat involved in developing the ability to walk, it appears, for the most part, that a "built-in" (genetically determined) engram (or structure) is being activated with progressive myelinization of motor areas and tracts.

What we are proposing is the exact analogue of this on the sensory side of the central nervous system. Few would question that when learning takes place, certainly *something* is happening in the brain. Perhaps most learning theorists will not question the proposal that learning involves setting up a model of reality based on real physical change. What is undoubtedly more controversial is our proposal that the sensory analogue to the *inherited* motor engram also exists. This is to say certain structures underlying cognition, certain models of reality, are genetically determined, some of which are probably as universally human as is the ability to "conceptualize" or "abstract." This is of major significance if true, since it would provide an explanation for the deep structural similarities in various human societies noted by Levi-Strauss, Piaget, Chomsky, Slobin, and others, as well as for the universal similarities in unconscious production so well documented by Jung. Furthermore, it does not require the Neo-Platonic stance of Levi-Strauss to explain his findings or the "mysticism" that Jung is often accused of invoking.

The Evidence For Neurognosis

One may properly ask, at this point, what evidence there is for the existence of inherited models of reality based on specific neuroanatomical structures. The answer comes from many disciplines and has been gradually accumulating since shortly after the turn of the century. Although such a proposition has been, in the main, resisted by the scientific community, we feel that an impartial examination of the evidence available today forces even the most skeptical to take seriously the possibility that some models of reality are inherited.

The evidence began to accumulate with the vast clinical psychiatric experience and cross-cultural studies of Carl Gustav Jung. Because of his later preoccupation with "mystical phenomena," most scientists lost sight of the objective clinical base that caused him to postulate the existence of "archetypes." Furthermore, his early use of a Lamarckian mechanism (later repudiated) to explain how "archetypes" originated tended to discredit him within the scientific community. Nevertheless, his use of the concept of "universal archetypes" to explain common themes in comparative folklore is masterful and compelling. His inference of the existence of "archetypes" (or primitive inherited models of reality) is reasonable given the evidence he presents. What is not so reasonable is his initial explanation of their nature. At first he took a Neo-Platonic position similar to that of Levi-Strauss. Thus in his *Commentary on the Tibetan Book of the Dead* (Jung, 1958a, pp. 292-293), he states:

> Among these inherited psychic factors, there is a special class which is not confined either to family or to race. These are the universal dispositions of the mind, and they are to be understood as analogous to Plato's forms (eidola), in accordance with which the mind organizes its contents. One could also describe these forms as *categories* analogous to the logical categories which are always and everywhere present as the basic postulates of reason. Only, in the case of our "forms," we are not dealing with categories of reason, but categories of the *imagination*.
>
> The original structural components of the psyche are of no less surprising a uniformity than are those of the visible body. The archetypes are, so to speak, organs of the prerational psyche. They are eternally inherited forms and ideas which have at first no specific content. Their specific content only appears in the course of the individual's life, when personal experience is taken up in precisely these forms.

In general, we would agree with this formulation, with the exception of his appeal to Plato's *eidola*. However, in a paragraph in the *Commentary on the "Secret of the Golden Flower,"* he formulates the basis of these unconscious models of reality in terms we can more easily accept (Jung, 1958b, p. 308).

> This unconscious psyche, common to all mankind, does not consist merely of contents capable of becoming conscious, but of latent dispositions toward identical reactions. Thus the fact of the collective unconscious is simply the psychic expression of the identity of brain-structure irrespective of all racial differences. This explains the analogy, sometimes even identity, between various myth motifs and symbols, and the possibility of human understanding in general.

In many ways, this statement was ahead of its time. In the absence of any neurophysiological mechanisms upon which to base such a statement, Jung's conclusion was accepted as idle speculation. We contend, however, this particular speculation was, in fact, correct. It is not our purpose to defend the work of Jung. Indeed we strongly disagree with Jungian psychology on many important points. But with respect to the matter of "archetypes" we feel his evidence is highly suggestive.

Our next body of evidence supporting the idea of inherited models of reality comes from the work of Claude Levi-Strauss. This anthropologist has amassed a great body of data strongly indicating that institutions in various cultures, which may appear structurally dissimilar to superficial analysis, prove to be identical in terms of "structure." By structure he refers to the unconscious processes by which man orders sensory input and alters his environment. Thus "dual organization" and "marriage alliance" are manifestations of universal "structural" organization. So compelling is the evidence that Levi-Strauss presents that it is fair to say that his theory is the most influential single issue in contemporary anthropology. Indeed, much of the motivation to produce this volume arose from a concern to help resolve the controversy by explaining structural organization by a neuroanatomic-neurophysiological model.

Although Levi-Strauss absolutely disclaims any influence from Jung whatsoever, several of their conclusions are remarkably similar. Both maintain that the "primitive" mind orders things in terms of opposites. Levi-Strauss, whose theoretical orientation was strongly influenced by the L'Année Sociologique tradition in sociology and the Prague school of linguistics, calls this tendency "binary opposition." The works of both men clearly indicate that they are talking about much the same thing. Furthermore, both men arrive at a kind of dialectic requiring some sort of reconciliation of the entities that the "primitive" mind necessarily polarizes. For Jung, this is primarily a phychological process that he terms *Conjunctio oppositorum.* For Levi-Strauss it is primarily the process of resolution suggested by Mauss in 1925 in *The Gift* (1967) and Hertz in *Death and the Right Hand* (1909; see also Needham, 1973). The social process, however, rests upon the initial mental process of binary opposition and its subsequent projection in the structure of social institutions.

The third body of evidence comes from theoretical linguistics. Since Chomsky published *Syntactic Structures* in 1957, followed by several other works, a controversy has been raging over the universality of the deep structure of language. Without attempting an explanation of the subtleties of transformational grammar, we can at least briefly describe the issues. Prior to Chomsky's revolutionary work, it was generally accepted in American (Bloomfieldian) linguistics that the structure of any language was arbitrary and the structural analysis of any language was unique to that language. To oversimplify somewhat, Chomsky and his followers, borrowing from the tradition of de Saussure, began to distinguish between the superficial structure and the deep structure of language. The superficial structure is equated with what was traditionally analyzed by linguists and is often thought of as being on the phonetic level. In other words, the sequence of sounds, intonational patterns, etc., comprise the surface structure. Deep structure is often thought of as being on the syntactic and semantic level. The difference between the active and passive voice is a frequently used and simple example. Thus, The boy broke the window and The window was broken by the boy are two sentences with dissimilar surface structure but similar deep structure. Deep structural analyses of many altogether unrelated languages have revealed marked similarities. This has led to the conclusion that linguistic universals do exist at least at the level of the deep structure of language. Lenneberg (1967) has given much compelling evidence supporting this view, so that it is probably not an exaggeration to state that the argument for linguistic universals at the level of deep structure has become accepted by the majority of theoretical linguists over the last few years. We shall return to this topic in greater detail in chapter 7.

A fourth source of evidence, that there are inherited structures by which input is processed and subsequent behavior organized, comes from ethology. Since most of this evidence comes from the study of insects and birds, it has been ignored until recently by the students of human behavior. It is common knowledge that many insects, exposed to a given stimulus, will immediately respond with a stereotyped behavior. It appears that the stimulus, "recognition," and motor response from a continuum of neural activity are organized in a highly structured fashion. The "pure instinct" of insects, some may claim, is too far removed from human behavior to be relevant. However, many species of birds show much the same behavior. Emlen (1970b) has shown the importance of celestial rotation in the development of migratory orientation. Such orientation is learned *much more rapidly* than is possible according to the laws of learning formulated by experimental psychologists up to a few years ago. The phenomenon is commonly mentioned in the ethological literature but was generally not considered by students of human behavior until it was noted by psychologists under controlled laboratory conditions.

The ontogeny of vocalizations among birds offers a further example of this phenomenon. R.A. Hinde (1969, p. 23) has noted for passerine birds that the "initial phase of song-learning occurs before the bird itself begins to sing. This first phase therefore fits neither the paradigm of classical conditioning, which involves the attachment of a response to a new stimulus, nor that of operant conditioning, which involves a change in the frequency of a response as a consequence of reinforcement." We will have more to say relative to animal communication in chapter 7.

These considerations lead us quite naturally to the fifth and perhaps most compelling body of evidence, i.e., that discovered by learning theorists themselves in experimental situations. In 1968 Brown and Jenkins demonstrated that pigeons learn to peck a lighted key that is paired with grain, even though pecking the key has no effect on grain. Although somewhat startling in itself in terms of the classical laws of learning, this result failed to produce full appreciation of its significance. This had to await the publication the following year of a paper by Williams and Williams (1969) entitled "Auto-Maintenance in the Pigeon: Sustained Pecking Despite Contingent Non-Reinforcement." This paper reported the results of four experiments conducted as a follow-up to those of Brown and Jenkins. The first showed that pecking was sustained even when pecks turned off the key and thus prevented reinforcement. The second experiment controlled for the possible effects of stimulus change and generalization. The remaining two experiments explored procedures that manipulated the tendency to peck the negatively correlated key by introducing other response keys that had no scheduled consequences. We feel that the general discussion of these experiments deserves to be quoted at some length since it considers the difficulties encountered in explaining the results of these experiments either in terms of classical (Pavlovian) conditioning or of instrumental conditioning. Furthermore, it is clear in this discussion that the pecking behavior becomes established and maintained by stimulus-reinforcer relationships, independent of explicit or adventitious contingencies between response and reinforcer. This is simply a technical way of saying that there is something about the stimulus (lighted key) in and of itself that reinforces the pecking behavior whether or not food is dispensed. The conclusion is that there is a built-in mechanism that markedly facilitates learning when a lighted key (or something similar in size, shape, etc.) is the stimulus. In their general discussion, Williams and Williams (1969: 518-520) state:

> Successful auto-maintenance—persistent, directed key pecking despite contingent nonreward—does not seem to be a natural implication of either operant or respondent principles. Because actual pecking of the negative key produces nonreinforcement, it cannot be directly maintained by adventitious relationships. It is supposed that unobserved behaviors preceding the actual striking of the key are adventitiously maintained, then either (a) one must assume that these adventitiously maintained precursors are inflexibly linked to pecking and account for why they are not extinguished or replaced by precursors for other responses, or (b) one must suppose that the precursors do not invariably lead to pecking but can precede other behaviors as well. If the latter assumption is made, then it is difficult to see why, by operant

principles, the precursors do not become closely linked to some other behavior because of the consistent reinforcement such linkage would produce. Even if attention—heightened and directed by adventitious reinforcement and giving rise to behavior through a look-peck coupling —is assumed to be responsible for the initial pecks (Brown and Jenkins 1968), it is difficult to see why the continual extinction of a specific response component would not lead to an operant shift in attention and inconsequential responding (when the negative key comes on, look at the continuous key and peck it). It is, for example, very difficult to see how the continuous key could ultimately control pecking but not compete with the negative key. It seems clear that stimulus-reinforcer relationships, and not only response-reinforcer interactions, play a special role in this phenomenon. Such a conclusion, of course, takes the phenomenon out of reach of a standard operant analysis, where the influence of stimuli depends on their discriminative function with regard to experimental contingencies.

Similarities to the respondent domain are easy to recognize: indeed, the auto-maintenance procedure is formally identical to the "omission training" procedure of Sheffield (1965) except that a key peck substitutes for a saliva drop. In the present case, the response at issue is topographically similar to the response made to food. However, even if one ignores the usual application of respondent analyses to autonomic rather than skeletal behavior, the directed quality of the induced pecking does not follow naturally from respondent principles (see also Brown and Jenkins, 1968). It is unclear, for example, why pecking would be directed at the key rather than the feeder, or indeed why it would be directed anywhere at all. Although respondent laws (which deal with laws of stimulus-reinforcer pairings) are no doubt pertinent to the present phenomenon, a detailed account of this phenomenon would seem to demand a serious augmentation of respondent principles to account for the directed quality of the response.

The most direct empirical precedent for the present phenomena is provided by the work of Breland and Breland (1961), who found, in several species, that response patterns that were related to a reinforcer could "drift" into a situation even though they delayed and interfered with actual production of the reinforcer. As in the present case, such behaviors were intrusive, counterproductive, and uncontrolled by their contingencies. In addition, it has been shown that pigeons trained to peck at other pigeons develop far more elaborate patterns of "aggressive" behavior than the actual contingency demands; the behavior is sometimes so vigorous that it continues into the period when reinforcement is available (Skinner, 1959; Reynolds, Catania, and Skinner, 1963; Azrin and Hutchinson, 1967). Similar effects have been reported in rats (Ulrich, Johnston, Richardson, and Wolff, 1963). When rats'

> running is reinforced, the speed at which they run is governed directly by the magnitude of the reinforcer, and fast running develops whether or not it produces more rapid reinforcement (Williams, 1966). These examples make it clear that contingencies of reinforcement alone do not determine when or how strongly some behaviors occur, even if the behaviors appear to be skeletal or "voluntary." As an instance of such direct control, the present phenomenon does not appear to be an isolated curiosity.
>
> . . .To relate the present work to the concept of "operant level" furnishes a context but does not provide an account. In further work, the concept of "arbitrariness," which is so frequently claimed for operants, will require close attention: is the action of reinforcement—direct or contingent—different when a response is "naturally" in the organism's repertory, or when it bears a special relationship to the reinforcer? More broadly, consideration should be given to ascertaining how frequently direct, as opposed to contingent, influences of reinforcers enter into the determination of "skeletal" or "voluntary" behavior in natural environments.

This last paragraph has led a number of investigators to reevaluate certain "anomalies" in learning experiments that had either been ignored, considered as error variance, or explained away by mechanisms more in keeping with orthodox learning theory. What has emerged is recognition that certain behaviors are very easily learned by some species, often even after only one trial, while they require the usual or "normal" number of trials to be learned by other species. Furthermore, those behaviors for which there is "preparedness" to learn seem to have marked survival value for that particular species. Thus it has been postulated that the similarity of the lighted key to a grain of corn or other cereal is sufficient to activate the prepared learning mechanism once it has been paired with a real grain of corn on only one or two occasions. Thus the pecking of the key is very rapidly established and is maintained over very long periods of time even when no food is forthcoming. Considering the diet required by a pigeon to survive, the existence of neural structures that markedly facilitate learning in this area is of obvious evolutionary significance for the species.

Treating species specific preparedness for learning in an evolutionary context, Seligman (1971) writes:

> Many readers have probably acquired some taste aversion after

> being sick to their stomachs. Garcia and associates . . . have found that this is a prepared form of classical conditioning. Rats learn to associate tastes, rather than external cues like lights, with nausea, and they can learn this in one trial even with a several hour delay between the taste and the illness. Pairing a light with food shock, on the other hand, takes several trials to acquire and can bridge a delay of only a few seconds. Note that this prepared learning reflects a real contingency which rodents have faced through the course of evolution: tastes are paired with poisoning, and the effects of poisons do not usually begin immediately. Such prepared learning gives a selective advantage.
>
> In the realm of instrumental learning, Brown and Jenkins (1968) shows that pigeons learn to peck a lighted key which is paired with grain, even though pecking the key has no effect on grain. Yet rats learn only by trial and error to press a bar for food and only if bar pressing produces food. In avoidance learning, birds have a great deal of trouble learning to peck a key to prevent shock, but can learn to hop up (Emlen, 1970a) or fly away (Bedford and Anger, 1967) to avoid shock. Rats have trouble learning to bar press to avoid, but learn to jump up to avoid in one trial (Baum, 1966). In discrimination learning, dogs learn readily to go to the left rather than to the right, if the cues which tell them which way to go are in different places, but can't learn if the cues differ in quality, rather than place. Conversely they can learn to put their paw up or keep it down if the cues differ in quality, but not of they differ in location (Dobrzecka, Szwejkowska, and Konorski, 1966). Seligman (1970) and Seligman and Hager [1972] discuss many other examples of prepared, unprepared, and contraprepared learning. The upshot of these examples is that learning itself may be quite different depending on how prepared the organism is for the particular contingency he confronts. (pp. 313-314)

One might ask at this point for a concise definition of the "preparedness" that we have been considering in the last few pages and what its relation is both to instinct and to accepted learning theory. In another work Seligman (1970, p. 408) once again gives a clarifying insight into the problem from a behavioral perspective:

> Confront an organism with a CS [conditioned stimulus] paired with UCS [unconditioned stimulus] or with a response which produces an outcome. Depending on the specifics, the organism can be either prepared, unprepared, or contraprepared for learning about the events. The relative preparedness of an organism for learning about a situation is defined by the amount of input (e.g., numbers of trials, pairings, bits of information, etc.), which occur before that output (responses, acts, repertoire, etc.), which is construed as evidence of acquisition, reliably

occurs. It does not matter how input or output are specified, as long as that specification can be used consistently for all points on the continuum. Thus, using the preparedness dimension is independent of whether one happens to be an S-R theorist, a cognitive theorist, an information processing theorist, an ethologist, or what have you. Let us illustrate how one can place an experimental situation at various points on the continuum for classical conditioning. If the organism makes the indicated response consistently from the very first presentation of the CS on, such "learning" represents a clear case of instinctive responding, the extreme of the prepared end of the dimension. If the organism makes the response consistently after only a few pairings, it is somewhat prepared. If the response emerges only after many pairings, the organism is unprepared. If acquisition occurs only after very many pairings or does not occur at all, the organism is said to be contra-prepared. The number of pairings is the measure that makes the dimension a continuum, and implicit in this dimension is the notion that "learning" and "instinct" are continuous. Typically, ethologists have studied situations from the prepared side of the dimension, while general process learning theorists have largely restricted themselves to the unprepared region. The contraprepared part of the dimension has been largely uninvestigated, or at least unpublished. (p. 408)

From a biogenetic structural point of view, "preparedness" would seem to imply the existence of specific neural structures that, once activated, determine the alignment of the organism to his environment both cognitively (insofar as this may be present in less developed organisms) and certainly behaviorally.

Seligman (1971) goes on to present strong evidence that many, if not most, phobias in man represent prepared learning and indeed have evolutionary significance. We shall consider this topic in more detail at the end of this chapter and in chapter 8.

Seligman then considers the possible significance of "symbolism" in terms of preparedness. He states (1971, pp. 317-318):

So, for a biologically oriented learning theorist, to what can the notion of symbolism amount? A is symbolic of B, if and only if human beings are prepared, in the sense defined, to learn that A is associated with B. If humans can acquire with A the properties of B after only minimal input, then it is meaningful to say that A is symbolic of B.

Even more speculatively, does preparedness range beyond simple symbolic associations? Are there ways of thinking in which humans are particularly prepared to engage, as Lenneberg (1967) has argued for language and cognition? If association, causal inference, and forms of cognition are prepared, are there stories that man is prepared to formulate and accept? If so, a meaningful version of the racial unconscious lurks close behind.

This last sentence is rather startling coming from a learning theorist. But it does indicate that biological science may be at the threshold of being able to offer a solid empirical explanation for the legitimate findings of such diverse scholars as Jung and Levi-Strauss. Awareness of this trend is already evident in the work of A.F.C. Wallace, who writes:

> If, as both human intuition and laboratory experiment suggest, reinforcement is *not* necessary for all learning, knowing, and thinking, then even where reinforcement *is* at work, there may be another process that contributes to the formation and maintenance of schemata [neural models]. Thus the . . . question is: what processes are there in addition to S-R learning that might contribute to the formation, maintenance, and application of schemata? Two sorts of answers suggest themselves, which are not mutually exclusive, and both of them of traditional interest to anthropologists: first, there may be certain cognitive capacities for forming kinds of schemata that are neurologically "wired into" all human beings, are genetically determined, and are presumably a product of evolution; second, there is a phenomenon that might be called "autistic thought" (without implying necessary psychopathology) that is responsible for reorganizing what has been learned, constantly recombining and differentiating elements in novel arrangements, going beyond information given, solving contradictions, and sometimes innovating new schemata. (1970, p. 78)

Evidence from all the above areas, "depth" psychology, structural anthropology, linguistics, ethology, and learning theory, all points in the same direction, i.e., that some of our models of reality are not learned according to the classical definitions of learning. A few at least—and perhaps the most important few, from the point of view of human adaptation—are based on genetically determined, neural models and structures. It seems a reasonable inference from all of this that the *content* of neurognostic models is first derived from phenotypic patterning based on instructions coded in DNA, and later modified by the organism's individual experience. Furthermore, it appears there may be particular latent neurognostic models that require specific kinds of stimuli in order to be activated. Finally, although some specific elements of subjective consciousness are derived from start to finish through individual experience, we still contend that the formal relationship of these elements as well is determined by neurognostic structuring.

Neurognosis and Adaptation

Let us now consider a few of what we consider basic adaptive functions of neurognosis and neurognostic structure. We certainly do not claim that the list is exhaustive, and we shall consider the structures related to language acquisition in a separate chapter.

In another work, one of the authors (d'Aquili, 1972) presented evidence for what he called the "cognitive imperative." The basic thrust of the argument is that all higher organisms have, as it were, a need to organize sensory input in a fashion meaningful and adaptive to the organism. The phrase "cognitive imperative" was used (*à la* the Scriven theorem) instead of "cognitive need" to distinguish the cognitive drive from the intensely emotion-laden "basic" drives such as hunger or sex. Consideration of the universal phenomenon of fear of the unknown was interpreted as the subjective reaction to frustration of this imperative. In man in particular, it was postulated, this imperative is particularly strong, since cognition is man's primary adaptive mechanism. Katz (1973) develops a similar argument from a somewhat different perspective in an analysis of the concept of purpose in the context of evolution. Empirical evidence supporting man's universal compulsion to order chaotic stimuli into meaningful patterns has most recently been presented by Adler and Hammett (1973a, 1973b) and Katz and d'Aquili (in preparation). The latter work analyzed responses to the mild 1973 Philadelphia earthquake. The primacy of a cognitive analytical response to the unidentified stimulus (mild earthquake) over an affective or other response was unquestionable. As pointed out in chapter 3, the cognitive imperative is ultimately responsible for abstract problem solving, tool using, and other activities of a general nature that are highly adaptive in any environment. What we are presenting here are a few of the basic ways in which information is processed. We contend that these modalities underlie both "psychological" and "cultural" universals, and that they account for them whether these universals are overt (similarity of content and structure) or covert (similarity only of basic structure). Thus phychological and/or cultural structures may be similar because they are generated from universal neurognostic structures.

First of all, and most simply, let us consider the ability to organize exteroceptive and proprioceptive neural stimuli into discrete groups. This serves as the basis of the "self" as a locus of apperception, on the one hand, and of the "other" on the other hand. As mentioned in chapter 3, part of this discriminating function is performed by neural connections from the anterior frontal cortex to the thalamus. Although extremely complicated, the connections involved for this primitive function are probably relatively simple compared to the extremely complex "holographic" model formation in the sensory association areas or in the inferior parietal lobule. Few anthropologists would deny that the self-other dichotomy is universal, and few neurophysiologists would deny that it is based on real neural structure. Furthermore, we suspect that few would doubt that this structure is genetically determined. Now the self-other dichotomy is a very important model of reality—probably one of the most fundamental "binary oppositions." When this structure does not function properly, the boundaries of "self" are lost with resultant psychotic symptoms, such as auditory and visual hallucinations. In other words, qualities of external objects or events appear to be "unreal," and internal events are projected outward, resulting in hearing voices that are not there, etc.

What is true of this simple neurognostic structure is also true of higher-order models, although, in our present state of ignorance, we are not always able to delineate the precise structure involved. One important neurognostic structure that we may discuss anatomically, to some extent, is the primitive, but universal, tendency to order reality into pairs that are usually subjectively experienced as opposites. As noted above, the process of binary opposition is very important in Jungian psychology and is often an essential part of dream structure. Also, Levi-Strauss has placed this cognitive tendency at the very foundation of marriage alliance and other essential human social institutions. It is of more than passing interest in this regard that lesions of the area comprising the inferior parietal lobule result in a loss of the ability to utilize basic logical-grammatical categories (Luria, 1966). These involve the dichotomous relationships "in-out," "high-low," "over-under," and "up-down," as well as a number of others. It would, therefore, seem, empirically, that one of the primitive functions of the inferior parietal lobule is to organize neural data into antinomies based originally upon primary spatial relationships. The basis of the antinomous structure of dreams, primary process (nonrational) thinking, mythological themes, and some basic social structures may, therefore, rest on the evolution of the inferior parietal lobule. Thus the organization of elements of neural input into semantically antinomous entities, enabled pretransitional hominids to relate one object to another, if only by their polarity. This would have effectively broken the stream of consciousness and served as the basis for the development of an understanding of relationships in general between discrete entities in the external environment. The tremendous importance of the phenomenon of antinomous thought can be appreciated only by those familiar with the literature of structural anthropology. It is not our intention to establish that importance here. We accept it as a given, but explain it as one of the basic inherited characteristics in the functioning of neurognostic models by which information from the external environment is processed.

Another neurognostic structure that appears to be genetically based is the ordering of sensory material into a causal matrix. This process, as we have indicated, arose from the cognitive extension of prehension, and the cognitive imperative is most keenly felt in the operation of this structure. It appears that man simply cannot function without ordering already integrated sensory input into causal sequences. The ability of a causal sequence to "explain" an event appears to allay man's primordial fear of the unknown. Furthermore, if a causal sequence is validly inferred from the environment (i.e., if the neural structure underlying the subjective sensation of causality operates only on external input and does not introduce higher order mental constructs into the causal chain) the neurognostic model enables man to gain some real mastery over his environment. It is one of man's most powerful faculties and, when operating only on environmental input, allows man to make valid predictions of consequent events given antecedent events.

There is some evidence (Premack and Premack, 1972) that chimpanzees think causally beyond the bounds of common sense under certain conditions. Consequently, we must date the cognitive extension of prehension to neural reorganization that took place in man's immediate prehominid ancestor. But if we can assume the chimpanzee to approximate this animal neurologically, we can conclude that the "drive" to order things causally was not very strong at the time of the hominid split. Undoubtedly the tendency to think causally (i.e., the neurognostic structure ordering input causally) was strongly selected for during the entire evolution of the hominid line. It became so strong that it has become the pivot of *Homo sapiens'* cognitive functioning. Indeed the strong need to explain things in terms of efficient causality (in the Aristotelian sense) can often undercut the practical value of this model of reality. Thus when a strip of reality (string of events) is presented to man without an initial terminus (first cause), man immediately fills the "defect" by elaborating a mental construct. Thus "powers," spirits, gods, etc, come into being, partly at least, to satisfy the cognitive imperative by supplying first causes to strips of observed reality. In this sense gods, spirits, and "powers" are an integral part of the causal neurognostic model of reality as it has presently evolved. Modern men of science are no exception to this rule. The ability deliberately to refrain from postulating a first cause for a sequence of events until it can be discovered by careful investigation is one of the glories of Western science. Yet it is a tenuous ability, operating only within the construct of scientific inquiry. It pervades only a small part of the cognitive functioning of scientists themselves, if only in the dreams and nameless fears, of even the most rational of rationalists. Furthermore, even within the realm of scientific investigation, jumps are often made to causes that cannot be physically demonstrated. The current astronomical concept of a "black hole" or Dirac's concept of negative energy states (in Duquesne, 1960) may have more of a logical basis than, say, the concept of a water spirit, but the inferential processes in both cases are formally identical. In other words, the causal neurognostic model of reality is functioning in both cases, but with considerable learned modification in the case of scientific inquiry.

Another adaptive function of the capacity for causal sequencing is in the structuring of social interaction. One of the authors (Laughlin, 1974a, 1974b; Laughlin and Laughlin, 1972) has demonstrated for one society in East Africa (the So) the tendency to structure interaction during "normal" environmental conditions so that the structure may easily change in response to predictable, cyclical deprivation. Laughlin generalized his findings to socioeconomic structuring in all societies that are faced with fitting social structure to fluctuating environmental resources:

> The primary biological function of any social system is the maintenance of a genetically viable reproductive population which is capable of perpetuating itself through time. This may be called society's *adaptive* function. . . . In order to fulfill its adaptive function, a society must assure the existence of modes of production and distribution of basic goods and services to at least the biologically critical fraction of its population. The "critical fraction" will usually consist of a sufficient number of its reproductively fertile young adults. (1974b)

Our point here is that the optimal value of the adaptive capacity of social organization would be unobtainable in the absence of causal sequencing. With this capacity, a human society can "prepare" itself far in advance of cyclical disaster or deprivation (e.g., drought, typhoon, war) so that when the event occurs, the population, at least to a biologically minimal extent, survives.

Let us consider briefly what is meant by causality to see if we can make some inference as to the specific neuroanatomical structure that is involved in this neurognostic structure of reality.

First of all, causality involves the probability expectation of a particular sequence. If we reason that, given A, we have a high expectation that B will follow from past experience, we say there is a causal relationship between A and B. Now when A and B are specific events immediately present in the sensory field of a given sense modality, there is a type of causal relationship that is prehended by many animals. An animal that can solve any kind of problem at all is capable, presumably, of ordering events in expected sequence provided the events occur within the sensory field of a single sense modality. As noted in chapter 3, this is a function of the temporal and spatial ordering of percepts by the frontal lobes. Reciprocal connections between the anterior portions of the frontal lobes and each sensory association area are presumably responsible for these causal perceptions.

But we have been centering on causal prehension within the bounds of common sense. The critical type of causality so characteristic of human cognition is the prehension of causal relationships independent of immediate sensory field. We contend that the difference between the two kinds of causality does *not* lie in the relationship between A and B. Indeed, the same ordering relationship, the same probability expectations, exist between the terms (A and B) in both the "primitive" and the "higher" form of causal inference. What does account for the difference is the nature of the terms; i.e., the nature of A and B. Terms A and B have the status of specific events in a single sensory field for lower animals, have the status of, at least potential concepts for the chimpanzee, and are almost always concepts for *Homo sapiens*. The crucial issues in the development of the perception of "higher causality" was, first of all, the evolution of the inferior parietal lobule and, second, the evolution of reciprocal connections between the anterior convexity of the frontal lobes and the inferior parietal lobule. Consequently, we can infer that the neurognostic structure governing causality involves the reciprocal connections between these two critical areas.

There are three categories of neurognostic models of reality that we contend are inherited, but we are at present unable to make any conjecture about the part or interrelated parts of the brain that make up their structural base. Because all three are of tremendous evolutionary significance, and by definition are universal, we feel they are not a result of conditioning, although, like all inherited neurognostic structures, they may be modified by learning procedures. All three involve an alignment of the subjective experience of "self" with some elements, or cluster of elements, in the external world.

First among these models is the alignment of self with conspecifics to form groups. It is, essentially, the basis of social behavior among both animals and humans. The importance of some sort of group structure to the survival of higher primates has been emphasized by many investigators (e.g., DeVore, 1965; Chance and Jolly, 1970). In another work one of the authors (d'Aquili, 1972) has delineated not only the importance of the primate group for survival but also its essential place in the development of higher cognition and "culture."

It has been suggested that the tendency to form groups of conspecifics is an extension of the mother-infant bond (see Radcliffe-Brown, 1924; Homans, and Schneider, 1955). In a certain limited sense, this may be true. We are aware, for example, of some strong evidence based upon lengthy observations of the Punta Santiago rhesus troop that there may exist, at least in some troops, a tendency toward sexual avoidance (incest taboo?) of the female genitor (Sade, 1968). We are also aware of the results of the classic experiments by Harlow and his associates indicating that young primates raised in artificial conditions without the mother-infant bond are unable as adults to relate socially with conspecifics. In itself, however, there are neither obvious nor necessary grounds for arguing that an extension of the cathected "mother" should take place to an entire group. In the first place, the functional bond between members of primate groups does not appear to be terribly similar to the mother-infant bond. If such a widespread extension does take place, the nature of the bond is certainly much modified.

In the second place, other explanatory devices appear more compelling to us than the extension of the mother-infant bond. Although we are willing to concede that cathexis on mother may be the excitatory stimulus for activation of the neurognostic structure(s) responsible for positive alignment between "self" and conspecifics, the latter set of bonds seems more determined by structures operating in satisfaction of other adaptive requirements. Adaptation in social primates, including hominids, has been a cybernetic phenomenon. The crucial problem for any primate group is to acquire the maximum corporate effective action with the minimum of energy expenditure. The particular configuration that proves the most cybernetically effective depends on a number of environmental variables (Denham, 1971; Kummer, 1971) and may vary in both rigor of alignment with conspecifics and degree of hierarchial arrangement from minimal (e.g., Chimpanzee) to maximal (e.g., Hamadryas baboons). Thus, at this point, we would contend that in-and-of-itself the mother-infant bond is insufficient to account for pervasive social behavior found among higher primates and underlying all human social institutions. And most particularly it cannot account for the variety of patterns of social behavior in different populations of the same primate species (Jolly, 1972; Itani and Suzuki, 1967). The mother-infant cathexis, nonetheless, represents the behavioral manisfestation of at least one neurognostic model of the sort that determines alignment of "self" with *specific* external objects.

The second category of neurognostic models of reality in which we are unable to postulate a definite neural structure involves the alignment of "self" with the external world conceived of as a global entity. This category of model concerns the dimension of "control of events." It is comprised of at least two separate neurognostic structures. One represents a positive alignment of "self" toward external events. Its dominance is reflected by a sense of mastery or mild euphoria and results in constructive action. The second involves a negative alignment of self to the external world and is manifested by a sense of helplessness or, clinically, by depression. The evolutionary significance of the first structure is obvious since it motivates outgoing positive action that often results in real control over external events. The evolutionary significance of the second structure is not as obvious. One can postulate that it provides a built-in mechanism for dominant males to move down the group hierarchy once they become too old to be effective. It is tempting to propose a mechanism of prepared learning such that a few failure experiences (perhaps in the area of dominating other group members via agonistic displays, etc.) cause this structure to be activated. The alpha male then processes all input in terms of a negative alignment of himself to the environment. Subjectively, he may experience helplessness or "depression" resulting in withdrawal and his relatively peaceful replacement by other males in the dominant position. Certainly excessive violence surrounding the filling of dominant positions among both males and females would undercut the survival value of primate group structure. Some provision must be made for a relatively peaceful moving down the group hierarchy. Activation of this neurognostic structure may provide the smooth transition the group needs to survive.

In any case, the study of depression as a cognitive structure has made great strides (Beck, 1964). Usually it is assumed to be a structure that results from instrumental conditioning and that is activated by certain stimuli such as personal loss. On the basis of its universal character, however, its basic elements, and its ease of activation, we feel that it is more probable that it is a neurognostic model. We shall review the evidence for depression as a reflection of a cognitive structure in chapter 8.

This category of neurognostic model of reality forces us to consider the concept of latent neural structures. Sade (1973) has shown that certain behaviors properly belonging to a certain subgroup of a given primate species (males, females, juveniles, etc.) can occasionally be observed to occur quite spontaneously in another subgroup. Thus, for example, an unquestionably male vocalization may rarely be observed to be emitted by a female or by a juvenile. His research has led Sade to postulate that the neural mechanism for the whole repertory of each class of behavior is present in each member of the species. The mechanisms that are actually activated are usually those proper to the subgroup. Thus males, for example, are usually the only ones that emit male vocalizations. But since the whole repertory of behavior is potentially present in each individual, certain conditions may occasionally activate a neural mechanism that normally is latent in that subgroup.

In an exactly analogous manner we contend that the neurognostic structure governing a positive alignment of the self with external events is activated by success experiences in early childhood (ages 1 to 3 years). Normally, it is this structure that processes neural input in a healthy individual. Experience of intense personal loss, however, or even certain endogenous chemical factors, may activate the structure that governs a negative alignment to the world, thus engendering a clinical depression.

The final category of inherited neurognostic models of reality to be considered in this chapter involves the alignment of "self" to a *specific* external object. If the alignment to a specific object is negative, it is called a phobia; if it is positive, it is called a "soterion." Although these relationships may unquestionably be formed through classical and instrumental learning procedures, evidence has been presented (Seligman, 1971) that many of the most intractable phobias are learned much more rapidly than could possibly be expected by orthodox learning theory. They are often formed upon one exposure to the phobic object. Although fear reactions can be conditioned to electric plugs, chains, or almost any object, they are quantitatively and qualitatively different from the more intractable phobias (Seligman, 1971). The latter involve fear of heights, insects, snakes, and various other animals. It seems more than simply coincidental that "prepared" learning operates in the formation of phobias that appear to have real evolutionary significance. We are dealing here with inherited structures governing a negative alignment of the self to an object that might have threatened survival in man's remote past. We will return to the nature of these phobias in chapter 8. What is perhaps more profitable to consider at this point is the probable inherited basis of certain "soteria." We have already alluded to the mother-infant bond. Bowlby (1969) and others have offered compelling evidence that this structure is activated in the mother by her infant's cries. Other evidence indicates that usually it is not capable of activation until 12 to 36 hours after birth. This may explain the strange phenomenon of the ease with which a mother may abandon her newborn, or give it up for adoption, so long as it is done within the first day or so after the birth.

Another example of an inherited structure governing positive alignment of self to a specific object is that governing sex objects. Activation of this structure appears to require not only exposure to the object but also the proper endocrinological matrix. Thus it is usually activated at puberty. In this regard, one may again invoke Sade's concept of latent neurological mechanisms to explain homosexuality. Presumably all individuals possess the neurognostic structure relating them sexually to the opposite sex as well as the structure relating them to the same sex. Normally the latter remains latent. Homosexuality occurs when, whether because of environmental, developmental, or endocrinological anomalies (or all three), the normally latent structure becomes activated and the normally activated one is latent. It seems that homosexuality, at least as manifestation of latent process and endocrinal anomaly, is universal in human societies. The precise fashion in which homosexuality will be manifest "culturally" will depend upon the predominant reaction of members of any particular society to resultant behavior. The continuum, of course, may vary from overt institutionalization and acceptance of homosexuality as among the Crow and Sioux forms of Berdaech to extreme negative sanction among societies that allow only covert or symbolic role shifts. Examples of the latter sort might be societies practicing the couvade, the ritualized acting out of the birth sequence by the male, a point similarly made by Charles Ackerman (personal communication).

Throughout this chapter we have placed heavy emphasis on the *inherited* quality of neurognostic structures. This is because this concept has great explanatory power in analyzing cultural universals, whether they are overt or whether they are manifestations of covert "deep structure" in Levi-Strauss' sense. Furthermore, the concept of inherited neurognostic structures simplifies the understanding of several human behaviors and feeling states, the study of which is properly the domain of psychology and psychiatry. This concept serves as a bridge between anthropology and sociology on the one hand, and psychology and psychiatry on the other. It attempts to explain what is both "culturally" and "psychologically" universal without invoking some sort of Neo-Platonic idealism. It rests on the neuroanatomic and neurophysiologic identity of the human central nervous system.

We do not for a moment deny the importance of the study of "culture," that is, the study of behavioral variation among discrete human populations. Much of cultural anthropology, however, is concerned with patterns of behavior resulting from neural models that although they are most likely holographic in structure, are *not* biogenetically based, i.e., do not contain neurognosis. We will discuss the question of the explanatory value of the concept "culture" in the conclusion of this book. The following chapter will explore in greater detail the relationship between one set of neurognostic structures and their extension into cognitive awareness.

References

Adler, H.M., and V.B.O. Hammett. 1973a. "The Doctor Patient Relationship Revisited, and Analysis of the Placebo Effect." *Annals of Internal Medicine,* 78:595.

——1973b. "Crisis Conversion and Cult Formation, an Examination of a Common Psychosocial Sequence." *Journal American Psychiatric Association.*

d'Aquili, Eugene G. 1972. "The Biopsychological Determinants of Culture." Reading: Addison-Wesley Modular Publications.

Azrin, N.H., and R.R. Hutchinson. 1967. "Conditioning of the Aggressive Behavior of Pigeons by a Fixed-Interval Schedule of Reinforcement." *Journal of the Experimental Analysis of Behavior,* 10:395-402.

Baum, M. 1966. "Rapid Extinction of an Avoidance Response Following a Period of Response Prevention." *Psychological Reports,* 18:59-64.

Beck, A.T. 1964. "Thinking and Depression." *Archives of General Psychiatry,* 10:561-571.

Bedford, J., and D. Anger. 1967. "Flight as an Avoidance Response in Pigeons." Paper presented at the meeting of the Psychonomic Society, St. Louis, Mo., October.

Bowlby, J. 1969. *Attachment and Loss,* Vol. 1. New York: Basic Books.

Breland, K., and M. Breland. 1961. "The Misbehavior of Organisms." *American Psychologist,* 16:681-684.

Brown, P., and H. Jenkins. 1968. "Autoshaping of the Pigeon's Key Peck." *Journal of the Experimental Analysis of Behavior,* 11:1-8.

Chance, N., and C. Jolly. 1970. *Social Groups of Monkeys, Apes, and Men.* New York: Dutton.

Chomsky, N. 1957. *Syntactic Structures.* The Hague: Mouton.

Denham, W.W. 1971. "Energy Relations and Basic Properties of Primate Social Organization." *American Anthropologist,* 73:77-95.

DeVore, I., and K.R.L. Hall. 1965. "Baboon Ecology" and "Baboon Social Behavior." In *Primate Behavior* (ed. by I. DeVore). New York: Holt, Rinehart and Winston.

Dobrzecka, C., G. Szwejkowska, and J. Konorski. 1966. "Qualitative versus Directional Cues in Two Forms of Differentiation." *Science,* 153:87-89.

Duquesne, M. 1960. *Matter and Antimatter.* New York: Harper.

Eccles, J.C. 1973. *The Understanding of the Brain.* New York: McGraw-Hill.

Emlen, S. 1970a. "The Influence of Magnetic Information on the Orientation of the Indigo Bunting (Passerina Cyanea)." *Animal Behavior,* 18:215-224.

——1970b. "Celestial Rotation: Its Importance in the Development of Migratory Orientation." *Science,* 170:1198-1201.

Hertz, Robert. 1909. *Death and the Right Hand* (1960 edition). Aberdeen: Cohen and West.

Hinde, R.A. 1969. *Bird Vocalizations.* Cambridge: Cambridge University Press.

Homans, G.C., and D.M. Schneider. 1955. *Marriage, Authority, and Final causes: A Study of Unilateral Cross-Cousin Marriage.* Glencoe, Illinois: Free Press.

Itani, J., and A. Suzuki. 1967. "The Social Unit or Chimpanzees." *Primates,* 8:355-381.

Jolly, A. 1972. *The Evolution of Primate Behavior.* New York: Macmillan.

Jung, C.G. 1958a. "Commentary on the Tibetan Book of the Dead." In *Psyche and Symbol* (ed. by V. de Laszlo). Garden City: Doubleday.

——1958b. "Commentary on 'The Secret of the Golden Flower.'" In *Psyche and Symbol* (ed. by V. de Laszlo). Garden City: Doubleday.

Katz, S.H. 1973. "Evolutionary Perspectives on Purpose and Man." *Zygon,* September/December.

Katz, S.H., E.G., d'Aquili, P. Gottesman, and J. Patys. "Cognition and Explanation: Responses to the Philadelphia Earthquake."

Kummer, H. 1971. *Primate Societies.* Chicago: Aldine.

Laughlin, C.D. 1974a. "Maximization, Marriage, and Residence among the So." *Canadian Review of Sociology and Anthropology,* 10(3):199-213. *American Ethnologist,* 1(1):129-141.

——1974b. "Deprivation and Reciprocity." *Man* (in press).

Laughlin, C.D., and E.R. Laughlin. 1972. "*Kenisan*: The Economic and Social Ramifications of the Ghost Cult among the So of Northeastern Uganda." *Africa,* XLII(1):9-20.

Lenneberg, E.H. 1967. *Biological Foundations of Language.* New York: John Wiley.

Luria, A.R. 1966. *Higher Cortical Functions in Man.* New York: Basic Books.

Mauss, M. 1967. *The Gift* (English edition). New York: Norton.

Needham, R. 1973. *Right and Left: Essays on Dual Symbolic Classification.* Chicago: University of Chicago Press.

Premack, A.J., and D. Premack. 1972. "Teaching Language to an Ape." *Scientific American* (October), 227(4).

Radcliffe-Brown, A.R. 1924. "The Mother's Brother in South Africa." *South African Journal of Science,* 21:542-555.

Reynolds, G.S., A.C. Catania, and B.F. Skinner. 1963. "Conditioned and Unconditioned Aggression in Pigeons." *Journal of the Experimental Analysis of Behavior,* 6:73-74.

Sade, D.S. 1968. "Inhibition of Son-Mother Mating among Free-Ranging Rhesus Monkeys." *Science and Psychoanalysis,* 12:18-38.

——1973. "Latent Processes and Biological Bases of Social Organization." Mb.

Seligman, M.E.P. 1970. "On the Generality of the Laws of Learning." *Psychological Review,* 77:406-418.

——1971. "Phobias and Preparedness." *Behavior Therapy,* 2:307-320.

Seligman, M.E.P. and J.L. Hager. 1972. *Biological Boundaries of Learning.* New York: Appleton-Century-Crofts.

Sheffield, F.D. 1965. "Relation between Classical Conditioning and Instrumental Learning." In *Classical Conditioning* (ed. by W.R. Prokasy). New York: Appleton-Century-Crofts.

Skinner, B.F. 1959. "An Experimental Analysis of Certain Emotions." *Journal of the Experimental Analysis of Behavior,* 2:264 (abstract).

Ulrich, R., M. Johnston, J. Richardson, and P. Wolff. 1963. "The Operant Conditioning of Fighting Behavior in Rats." *Psychology Records,* 13:465-470.

Wallace, A.F.C. 1970. *Culture and Personality* (second edition). New York: Random House.
Williams, D.R. 1966. "Relation between Response Amplitude and Reinforcement." *Journal of Experimental Psychology,* 71:634-641.
Williams, D.R., and H. Williams. 1969. "Auto-Maintenance in the Pigeon: Sustained Pecking Despite Contingent Nonreinforcement." *Journal of the Experimental Analysis of Behavior,* 12:511-520.

VI

Evolution and Empiricism

> A good theory, however, by its own definition, cannot remain intact. Every research based upon it, by virtue of the variance in the data which is not accounted for, should lead the investigator back to the theory to revise and elaborate and thus develop new hypotheses for test. It is this interplay between induction and deduction which contributes to the progress of a science, with theory functioning as an instrument of logic in the service of extending the possibilities for acquiring data.—SILVERMAN, 1964

IN THIS CHAPTER we wish to offer an empirical model of scientific theory-construction, a model that is really a logical adjunct to the biogenetic structural theory discussed in the last several chapters. We may term the present model *inductive-deductive alternation.* It is our contention in calling this an empirical model that all any philosopher of science is doing when he discusses how science progresses, or ought to progress, is constructing a model (verifiable or not, depending upon the philosopher) of empirical reality (see Kaplan, 1964, for discussion of "reconstructed logic").

Sciencing, or the doing of science, is a biologically given process, one that all of us do almost all the time and do with partial awareness at best. It is a real phenomenon, existing in the real world, and one about which men have reflected for ages. As such, it is perfectly amenable to direct or indirect empirical observation, theoretical formulation, and explanation. As with models pertaining to any nontrivial and complex phenomenon in the world, some models are better than others, better, that is, in the positivistic sense of offering an explanation from which accurate predictions about the phenomenon may be made.

What has usually inhibited an adequate examination and explanation of the process of sciencing has been the almost universal failure to recognize the process as real, rather than normative (a set of rules for correct inquiry, a system of empirical logic, a profession of all but magically powerful practitioners, a complex and special methodology, etc.). Philosophers of science often attack problems of scientific inquiry as though they were capable of some *à priori,* logical solution. Northrop (1947), for example, in discussing how the scientist makes certain that his theoretical formulations reflect relationships extant in the real world, has recourse to a discussion of the logic of "epistemic correlations," that is, the set of logical rules by which theoretical entities are assigned to real entities. From our point of view, it would be more profitable for the philosopher of science to observe closely the way science actually happens, rather than carry on armchair reflections after the manner of nineteenth century social philosophers. Piaget (1973) has stressed this point repeatedly in his work on the ontogeny of levels of cognition:

> What I maintain . . . is by no means particular to infant thought; it [the structure of reason] is found again not only in every adult thought but also in the development of scientific thought. Thus mathematicians of all times have reasoned by obeying, without knowing it, the laws of certain structures, the most exacting of which is the structure of the group, which is easy to see in the work of Euclid's *Elements,* for example. Mathematicians knew nothing of this, and it was only in the early nineteenth century that Galois "became aware" of the existence of such a structure now recognized by everyone as fundamental. Similarly, Aristotle in creating logic by concentrating on the manner in which he, as well as his contemporaries, reasoned, "became aware" of some of the simplest structures of the logic of category and of syllogism. What is very interesting, however, is that at the same time he was not aware of a whole set of structures which he himself used and which are the logic of relations: This also was not understood until the nineteenth century with Morgan's work, etc. (p. 34)

We will argue that science—all science, everywhere—progresses by a relatively well-disciplined application of inductive-deductive alternation. This is not to say that "science" always and everywhere progresses or must progress, or for that matter *should* progress. We are in partial agreement with Kuhn's (1970) explanation of why science very often does not progress. What is suggested here is that when it does manage to progress, new paradigms are formulated by an alternation between induction and deduction.

Sciencing may be construed as a continuum of intellectual involvement that may range between two extreme poles, "induction" being one pole and "deduction" being the other (see Skyrms, 1966, p. 12). When sciencing is done relatively self-consciously, what is being manifested is the extension into awareness of the empirical modification cycle, the empirical feedback system by means of which the brain continually revises its holographic models so they are capable of accurately predicting external events (see chapter 4). Men everywhere, in every society, do science. That is, they formulate models of reality suggested by perceived regularities in empirical data and test those models by the derivation of predictions that in turn are evaluated in terms of accuracy of fit with the real world.

More often than not, philosophers of science and anthropological theorists tend to take extreme positions on the question of the appropriate logic of inquiry. Some have taken the view that a discipline can progress as a science only when it reaches the capacity to generate deductive theory. At that point, the discipline no longer needs to rely upon the less prestigious body of mere empirical generalizations gained inductively from a specified data field. Levi-Strauss, for example, has often argued that anthropology can become a true science only when it begins to construct deductively the primary, and essentially unconscious, structures underlying apparent diversity in human society. In comparing the relative status of anthropology with that of linguistics as a science, Levi-Strauss (1967) has written:

> Among all social phenomena, language alone has thus far been studied in a manner which permits it to serve as the object of truly scientific analysis, allowing us to understand its formative process and to predict its mode of change. This results from modern researches into the problems of phonemics, which have reached beyond the superficial conscious and historical expression of linguistic phenomena to attain

> fundamental and objective realities consisting of systems of relations which are the products of unconscious thought processes. (p. 57)

The unconscious structures to which he refers may not be discovered by correlational analysis of data sets. Correlational regularities in a data set amount to what Levi-Strauss has called "statistical models," whereas the underlying structural model he terms "mechanical" (1967, pp. 275 ff.). The latter as manifest in any particular society represents a partially prehended, unconsciously applied rendition of a Platonically perfect model. And it is possible to know that Platonic model only by recourse to a deductive theoretical perspective. An adequate deductive formulation (for Levi-Strauss) is one that allows the predictive generation of any particular local structure through a process of logical transformations.

Another anthropological theorist who may be classed as a deductivist is Elman Service. In his work *Primitive Social Organization* (1962) Service proposed an evolutionary explanation of social diversity. At the core of his theory are four "pure" types of human social organization—the patrilocal band, the tribe, the chiefdom, and the state or archaic civilization. All societies, had they not been affected by modern industrial society, would be found in one of these categories. Contact with the West, however, has caused a measure of social perturbation in many societies and requires the modification of the "pure" types to include the "composit" band, as well as the "compound" and "composit" tribe. Each step up the evolutionary ladder was necessitated in each case by a shift in technological, ecological, and economic factors allowing for increased population and demographic centralization.

Pure deductivists like Levi-Strauss and Service are prone to generating *post hoc* and tautological propositions, which, as often as not, explain little or nothing. The problem lies in what Hempel (1966, pp. 72 ff.) has called the determination of "bridge principles." How does the analyst bridge the gap between the world of theory and the world of real phenomena, portions of which he wishes to explain? The pure deductivist is often either incapable of the effort or is unconcerned. Levi-Strauss has maintained, for example, that a correlational test of a deductively derived structural model is of no validity in the confirmation or disconfirmation of the empirical value of the model. The problem with Elman Service's construct, of course, is that it is *post hoc* with reference to the data from the ethnographic present, and simply untestable with reference to hominid evolution.

At the other extreme are what might be called the "pure" inductivists—theoreticians who hold that generalizations about society must be obtained from a data set. It is assumed that the universe is ordered and that order is knowable to science if scientists but collect sufficient data. Alfred North Whitehead probably best exemplifies this extreme in the philosophy of science. Whitehead posited what he called the fallacy of the "bifurcation of nature," the division of the universe into the world of reality and the world of science (Whitehead, 1957, p. 443; 1964, chapter 2). The fallacy is committed by theorists who insist upon positing the existence of theoretical entities (like the neutrino, mass, and dual organization) to explain observed phenomena. The order in the empirical world is evident enough to those who look for it without recourse to theoretical bifurcation.

Examples of anthropological inductivists abound—on both sides of the Atlantic. A.R. Radcliffe-Brown, throughout most of his writings at least, implied that the correct presentation of data on any particular society required a theory-free exposition of the structural-functional relations between aspects (subsystems) of the social system. Franz Boas urged his students to enter the field and collect masses of data that were to be later ordered by application of various typological methods. Even today a student of a student of Boas, Harold Driver, has brought a great deal of sophisticated reasoning and formal methodology to the acquisition and inductive ordering of empirical data.[1]

Pure inductivists in anthropology tend to render one of two results, or a mixture thereof. They either produce reams of apparently disassociated and often irrelevant data or they present covert

[1] It should be remembered that Driver has utilized statistical methods as well for the empirical testing of the theoretical formulations of others (see Driver, 1966).

theoretical formulations, embedded, as it were, in the concepts and typological structures utilized in the basic research.

A major pitfall in the application of either extreme is in the usually erroneous understanding of the epistemological status of concepts. As a number of philosophers have demonstrated, there are both a logical and a cognitive connection between theory and concept, and it really makes no sense to speak of "theory-free" concepts. Indeed, the philosopher of science, Paul Feyerabend (1965), views theories as the repositories of meaning in such a way that concepts deriving their separate meanings from different theories are incomparable. He further treats language as theory and words, therefore, as concepts positing a rather complicated Whorf-Sapir hypothesis[2] in the study of the philosophy of science. Although we would tend to argue with Feyerabend's basic assumptions (Feyerabend, 1965) *vis à vis* the origin of human empiricism, we are forced to applaud his emphasis upon the inextricable relationship between theory, concept, and observation. He speaks most poignantly to anthropologists of the inductive extreme who, at times, have contended they were presenting nontheory-laden field observations, when in truth their "theory," or set of theoretical assumptions, was implicit in the research design, words, and descriptive concepts used.

Advocates of either philosophical extreme—the pure inductive position as well as the pure deductive position—commit what Korzybski (1958) termed the Aristotelian fallacy, confusing the word with the thing signified. Practitioners of either persuasion often lose sight of the fact that the logical constructs under which they are operating are merely proposed as models of how man sciences. They offer the practicing scientist formulas for doing science only to the degree that they are accurate models of the real process of inquiry. Our contention is that the extreme position at either pole is really creating a severely truncated system, reflecting at best only half the total process. Too rigid adherence to the dictates of either extreme in anthropology results in scientific half jobs. Indeed, one often has

[2] The Whorf-Sapir hypothesis states that not only will a language be affected by the environment in which it is found but that the language will also, to some extent, affect the way in which the environment is perceived by speakers of that language.

the feeling that the analyst has stopped short of truly exciting conclusions or tests of conclusions.

Nowhere has this been more evident than in the remarkable controversy revolving around the cause of the notably close and affectionate relationship between a man and his mother's brother, and the inclination toward marriage with mother's brother's daughter, in many South African and other societies. When discussing his observations, Radcliffe-Brown (1924) attempted to explain the phenomenon as one by-product of a man's affection for his mother, that his sentiment for her is extended to cover her brother as well. Levi-Strauss, it may be remembered, took exception in 1949 to the conclusion offered by Radcliffe-Brown and suggested in turn that there is no logical reason for unilineal cross-cousin marriage (marriage between ego and his mother's brother's, or father's sister's, daughter) to be linked with directionality of descent (1969, pp. 99 ff.). Rather, the preference of one type of cousin marriage (namely matrilateral) over the other results from an inherent superiority of one type to create and sustain viable alliances between otherwise disparate groups.

Taking up the argument from an ostensibly psychological point of view, and in defense of Radcliffe-Brown, George Homans and David Schneider reasoned that because of the tendency in patrilineal societies for ego to transfer his affection from his mother to his mother's brother's entire group, a close correlation should be found among the world's societies between patrilineal descent and matrilateral cross-cousin marriage. They sought and found such a correlation in an availability sample of 33 societies (1955, p. 34).

In reaction to Homans and Schneider, Levi-Strauss (1967, p. 317) simply reiterated his contention that no causal relationship exists between type of marriage and type of descent and offered a *post hoc* explanation of Homan and Schneider's "statistical correlation." Rodney Needham (1962) entered the fray with a (sometimes caustic) defense of Levi-Strauss in which he (i) denied that the latter was including "preferential" (one type of marriage socially preferred, but not mandatory) with "prescriptive" (one type socially mandatory) marriage systems in his discussion of unilateral exchange, and (ii) charged Homans and Schneider with committing an unforgivable psychologism, and with having accused Levi-Strauss erroneously with reductionism (pp. 4, 24).

The final word in the controversial exchange was spoken by Levi-Strauss in the preface to the second edition of *Elementary Structures.* Here again he reiterates his previous position and adds the contention that Needham's defense presented an erroneous version of his position, that the preferential-prescriptive distinction was unfortunate as "the notions of prescriptive and preferential marriage are relative: a preferential system is prescriptive when envisaged at the model level; a prescriptive system must be preferential when envisaged at the level of reality. . . ." (1969, p. xxxiii). The fact that matrilateral cross-cousin marriage does not predominate in a statistical sense in a society is no indication that the structure (with which he is principally concerned) is not present: ". . . a system which recommends marriage with the mother's brother's daughter may be called prescriptive even if the rule is seldom observed, since it says what must be done. The question of how far and in what proportion the members of a given society respect the norm is very interesting, but a different question to that of where this society should properly be placed in a typology" (1969, p. xxxiii).

There are many interesting lessons to be gleaned from this protracted controversy. However, we present the case here to illustrate an incompatability that can emerge from a clash between extremes of inquiry. Homans and Schneider, largely representing the inductivist position, are primarily impressed with a statistical correlation between descent and type of marriage. That is, little effort is expended on their part to develop and refine a more sophisticated model than that presented by Radcliffe-Brown before them. Their real concern is to prove the structuralist's position wrong, and to do this they point to an empirical generalization they induced from a limited availability sample. Levi-Strauss, on the other hand, barely defends himself in the fray by a redundant reiteration of his original position. One gains the distinct feeling that he is all but intolerant of empirical tests of his deductive propositions. His models are sufficiently validated if they are logically consistent and are capable of transformations approximating any particular system abstracted from individual societies. The result of the extreme positions depicted here is that neither side contributes to the progressive inquiry of the other. In a word, they "talk past each other." Levi-Strauss devotes merely a footnote in answer to a booklength attack of his position by Homans and Schneider; he fails altogether to proceed beyond the formulation he had already and long before devised. Homans and Schneider, in contrast, fail utterly in understanding what Levi-Strauss means by the distinction between mechanical and statistical models, and hence they miss the chance of pyramiding their own formulation beyond that expressed by Radcliffe-Brown years before. It might also be noted that neither position is *in fact* a "pure" one. Both take initial cues from inductive regularities, both concern themselves, at least to some degree, with model building. Yet the cycle remains incomplete. Feedback is blocked to the models, and, thus, improvement is inhibited.

Inductive-Deductive Alternation

What the terms induction and deduction depict, then, are actually two phases of a continuum of inquiry, neither of which is meaningful in any actual sense when abstracted from the whole. Viewed in this way, it makes sense to term the continuum of inquiry itself inductive-deductive alternation. We may now turn to a more precise examination of the present model of sciencing.

A scientific model may be defined as a set of propositions that contain entities and relationships between entities, the existence of at least a portion of which is untestable in any direct sense in the real world. Although the direct untestability is a characteristic of models, they are necessary in order for men to relate objects and events in time and space that are beyond common-sense prehension. Thus enters a seeming paradox—at least at the level of "reconstructed" logic—that the scientific model purports to tell us something about the real world and yet is untestable by direct correspondence with the real world. The paradox evaporates, however, when we place the evolutionary status of empirical inquiry in its proper perspective. Scientific models may be seen as special kinds of neurognostic models that are extended into awareness, perhaps even objectified through written language, but that exhibit the same qualities as other models derived consciously and subconsciously through the empirical modification cycle.

Scientific models, once formulated, are then tested for accuracy of fit with the real world. This is done by generating hypotheses that, unlike the models, themselves, predict (or retrodict) something about the real world that *can* be directly verified. But how are these predictive hypotheses constructed, and how can we be sure they have the power to confirm their parent model? The answer is simple. A predictive hypothesis should be the conclusion of a valid deductive argument, the premises of which are all, or some, of the theoretical propositions constituting the parent model. Depicted schematically, this logical relationship between model and hypothesis would look something like Figure 6.1. Model propositions 1 through 4 become logical premises 1 through 4 from which we derive conclusion A by deduction. Conclusion A then becomes a deductive hypothesis and one that stands in a special logical relationship to its parent model. Because of its deductive origin, if premises 1 through 4 are all true, conclusion A must also be true. Thus, if we take conclusion A out into the real world and find that the event or relationship predicted therein is, in fact, *not* true, then we know there is something wrong with one or more of the original premises. We must use caution here, however. What does it mean if the event or relationship predicted in conclusion A proves to be true when examined in the real world? Does this mean that premises 1 through 4 must also be true? Emphatically not! The truth of conclusion A merely confirms that premises 1 through 4 *may* be true, and that there is no reason as yet to believe they are false. To put it formally, our premises are sufficient, but *never* necessary, conditions for conclusion A. Conclusion A could just as well have been derived from a totally different set of premises. This is a very important point, because it explains why it is usually necessary to test a scientific model, not once, but time and time again. With each succeeding test, we gain increasing confidence that the theoretical premises embedded in our model posit to some extent actual events in the real world.

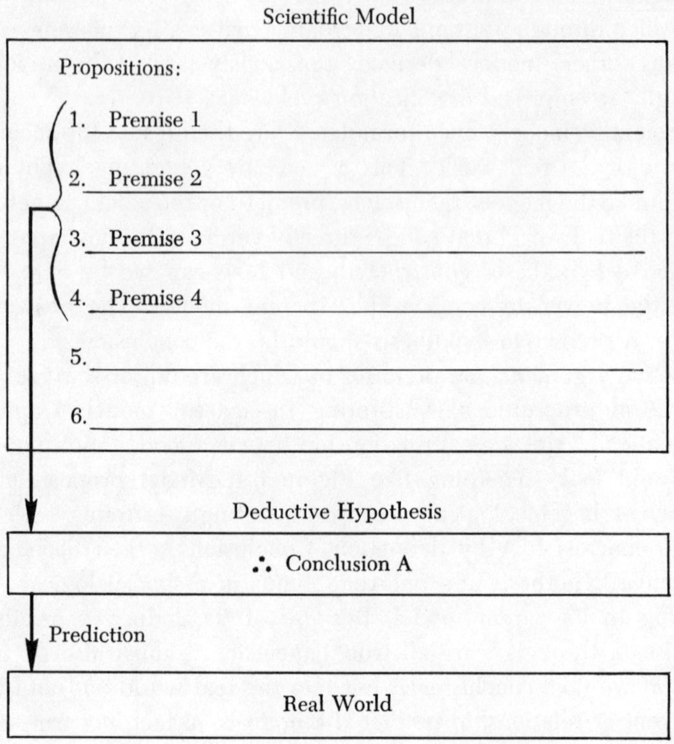

Figure 6.1. The Relationship Between Scientific Model and Deductive Hypothesis: The Deductive Phase.

But it is usually not as simple as that. One or more of the critical first hypotheses will, at best, only partially predict events in the real world. If the results of experiments and field observation are at variance with the predictions deduced from our model, progressive refinement of the model is necessitated. The variant results then become "inductive" input into the model—feedback drawn into the model through the hypothesis. The old hypothesis, conclusion A, is first modified so that it becomes a proposition, conclusion A', which, had it been originally deduced from our model, would have predicted the correct observational results. We must then return one step further up the inductive ladder to the propositions within our model that formed the premises from which conclusion A was originally deduced. We wish now to determine how these propositions may be changed so that, taken as a new set of premises, 1' through 4', the new conclusion A' might be deduced. This step might be termed the *post hoc* stage because if we left our model as complete at this point, we would be committing the teleological or *post hoc* logical fallacy in asserting that an event that exists in the real world had to exist.

Once we have modified our model so that it now includes propositions 1' through 4', and a *post hoc* deductive hypothesis, conclusion A', the process of inquiry and scientific model building shifts back to the deductive phase of the cycle. It is necessary to generate a new hypothesis, using all or some of our modified propositions as premises in a deductive argument, the conclusion of which, conclusion B, predicts some event or relationship in the real world that was not contained in the prediction entailed in the original conclusion A, or in modified conclusion A.'[3] Thus we are now independently testing the modified model and, at the same instant, avoiding the pitfalls of the *post hoc* fallacy. Figure 6.2 depicts the process of inquiry to this point.

The process continues as before with predictions made by conclusion B being measured for accuracy in the real world, empirical feedback perhaps necessitating modification of conclusion B to conclusion B', a second modification of initial propositions and finally, the deduction of a third deductive hypothesis, conclusion C. Thus by a lengthy series of alternations, between the deductive and the inductive poles of inquiry, a body of predictive confirmation is amassed in support of the model. We are not saying, of course, that all models succeed in being confirmed. What we are describing is an empirical theory of how successful scientific models are confirmed and inadequate ones rejected.

It might be well to reiterate that the process of scientific inquiry was not a by-product of Western philosophical tradition. It is rather a psychologically compartmentalized application of a proc-

[3] Conclusion A' may be validly tested, of course, on a body of data different from, and independent of, the original body of data. This procedure will also avoid the *post hoc* fallacy.

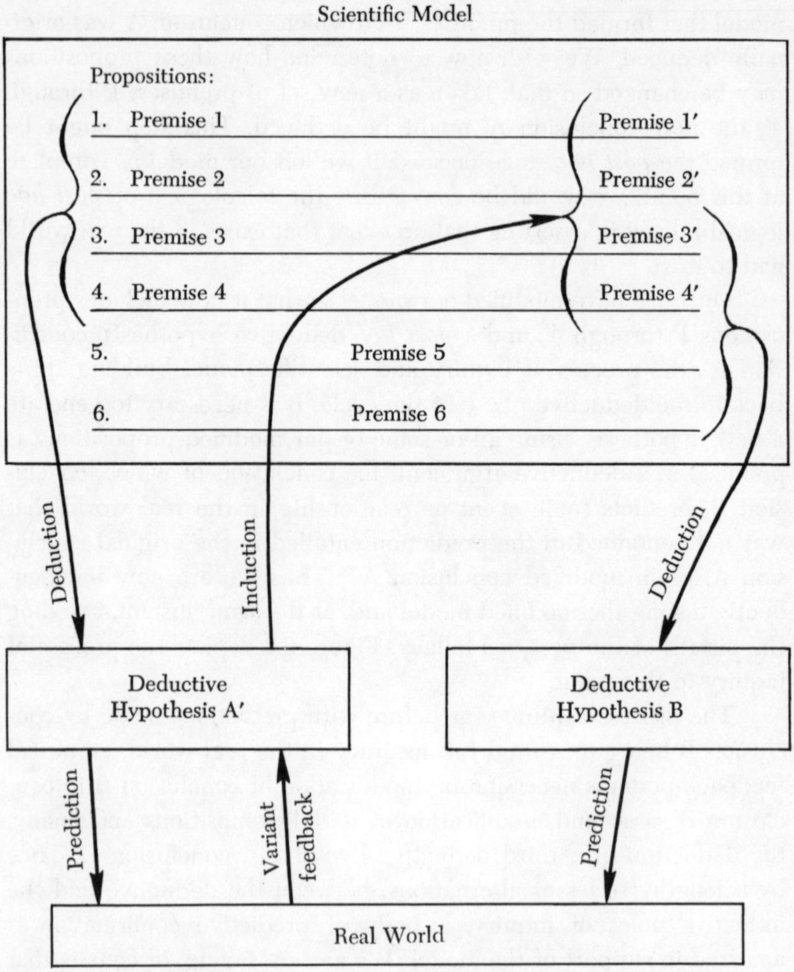

Figure 6.2. The Relationship Between Scientific Model and Deductive Hypothesis: First Inductive/Deductive Alternation and Second Deductive Phase.

ess—the empirical modification cycle discussed earlier—that is genetically inherent in our neurophysiological structure. Let us directly compare the phases of both the process of scientific model construction and the process of empirical modification of holographic models.

Empirical Modification Cycle	*Scientific Theory Construction*
1. Holographic model (perhaps in form of neurognosis)	1. Initial model formulation
2. Generation of expectation	2. Deduction of hypothesis
3. Behavior based upon expectation and inputting of perceived response	3. Hypothesis tested by prediction, observation, and accumulation of data set
4. Data stored in short-term memory	4. Storage and ordering of empirical data set
5. Expectation evaluated as to accuracy of fit with material in memory	5. Confirmation or rejection of hypothesis
6. Physical readjustment of holographic model	6. Modification of hypothesis and theoretical propositions
7. Generation of expectation	7. Deduction of hypothesis

Another point of interest about inductive-deductive alternation is that the cycle may be entered at any point. The individual scientist may begin with a model suggested by another theorist and move to test and further refine the model he has borrowed. In a sense, then, he is entering at the deductive phase in the cycle. More often the scientist will be confronted by a body of data exhibiting a regularity of patterning (perhaps discovered by accident), a regularity he wishes to explain. In order to do so, he develops or obtains a set of theoretical propositions that, when they become the premises in a deductive argument, logically entail an hypothesis that predicts the regularities with which he was concerned in the first place. In other words, he has entered the cycle at the inductive phase. Either way, the cycle must be accomplished for the generation of a complete empirical model.

Germane to our position on "awareness" is the fact that men are largely unaware of most of the associative processes occurring in their brains, especially those processes of primitive origin (neurognosis). Men are barely aware, for instance, of the constant process of inductive-deductive alternation (empirical modification cycle) by means of which their neural models are adaptively refined. On the rare occasions when we are aware of the process, or consciously direct the process toward the formulation of explicit models of reality, we call the result "rational," "empirical," and "science." Oddly enough, when observing the results of the same process among non-Western societies, anthropologists and others have traditionally characterized the process as "magic," "myth," "sorcery," "cosmology," "Weltanschaung," and the like. Paul Radin in 1927, however, understood this as "doublethink" when he wrote *Primitive Man as Philosopher* (1957). This conceptual error was later exposed by Levi-Strauss in *The Savage Mind* (1966). The most recent and certainly the most definitive exposé of the ethnocentrism of such anthropological terminology is Robin Horton's (1967).

Horton understands that all men approach the problem of model building, and the solution to the problem of "correspondences" between model and reality, in the same fashion. Horton also sees the relevance of the fact that we term models of reality "scientific theory" only if they relate entities or events beyond common sense (1967, pp. 53 ff.). In explaining why anthropologists have time and again perpetuated this conceptual error, Horton says:

> One reason, as I suggested before, is that many of them move only in the common-sense world of Western culture, and are unfamiliar with its various theoretical worlds. But perhaps familiarity with Western theoretical thinking is not by itself enough. *For a thoroughly unfamiliar idiom can still blind a man to a familiar form of thought.* (p. 52; emphasis added)

The salient point here is that all men, everywhere, theorize to some degree. Furthermore, men everywhere construct theoretical models in the same fashion and for the same reason—because they are genetically predetermined to do so.

To offer an instructive example from the field work of one of the authors, the So of Northeastern Uganda believe in the existence of an entity they call *tegwec*. The *tegwec* dwells in all mountain streams and pools and is ordinarily invisible to human detection. When a person bathes in the stream, he may be watched by a *tegwec,* though he cannot reciprocally watch the *tegwec*. The bather is in no danger unless the *tegwec* is angry for some reason. If angered, a *tegwec* is quite dangerous and will cause a bather later gradually to weaken and even to die. Ultimately, the family of the victim will seek the help of a practitioner (*ic*) who, when finding the patient has recently been bathing, may well diagnose the action of a *tegwec* and order a poolside curing ceremony.

The reason for this bit of ethnography is that the pools in and around this particular area are infected with a disease we choose to call bilharzia or "liver fluke." This is a disease that causes precisely the same symptoms the So associate with attacks by a *tegwec.* Both concepts are used to explain a body of empirical data—in this case infected persons—and both link the victim with prior contact with water and finally with a waterborne entity. That the concept of bilharzia, combined with the theory and practice of Western medical science, is capable of telling us more about the real world than is the theory of the *tegwec* (such as how to get rid of the infection from the pools, for instance) should not blind us to several important considerations: (i) Both concepts are embedded in models of the real world. (ii) Both were formulated empirically to explain a data set abstracted inductively from real events. (iii) Both models are used to predict (and retrodict) events of the same nature in the real world. For an excellent and detailed discussion of this issue the reader is referred to the section on beliefs (chapter 7) in Goodenough's *Cooperation in Change* (1963).

It should be stressed that though the underlying mode of human inquiry is structured biogenetically, the particular phenomena of concern, the questions asked about the real world, and, hence, the precise content of models usually are not. These kinds of considerations are largely involved with the values and social structure of the particular society. And yet the theoretical formulations of men everywhere are flavored with the same logico-structural spices. Nowhere is this more evident than in the universal commission of what we call "logical fallacies." Men everywhere generalize from the particular, appeal to authority, render *post hoc* formulations, and affirm the consequent. These are clear manifestations of the biogenetic structure of the empirical modification cycle.

The theory of biogenetic structuralism offers a scientific explanation for two heretofore knotty problems in the philosophy of science, the seemingly inherent tautological nature of deductive nomological models and Hume's problem of the verification of induction.

F. S. C. Northrop (1947, pp. 144 ff.) clearly outlined for us some years ago the inevitable fallacy involved in the confirmation of deductive scientific theories. In his terms, if we allow A to represent the deductive theory and B to represent hypotheses deduced from the theory for testing, the process of testing may be represented as follows:

(1) If A then B; B is not the case; therefore, A is not the case.
(2) If A then B; B is the case; therefore, A is the case.

> The logic of these two arguments has been fully and conclusively treated by logicians in their analysis of the hypothetical syllogism. This analysis shows that whereas the argument in the case of the falsification of a scientific theory is formally valid, that in the case of its confirmation is logically invalid, since it commits what logicians term the fallacy of affirming the consequent. We find ourselves, therefore, in this somewhat shocking situation: the method which natural science uses to check the postulationally prescribed theories in its more mature stage of development is absolutely trustworthy when the proposed theory is not confirmed and logically inconclusive when the theory is experimentally confirmed. (1947, p. 146)

As mentioned earlier in this chapter, the propositions of the deductive model can constitute only sufficient, but never necessary, conditions for their deductive hypotheses.

The problem raised by David Hume (1955, pp. 40 ff.) in the mid-eighteenth century, and which has concerned logicians to the present day (see Skyrms, 1966, pp. 22 ff.), is that of the valid confirmation of conclusions reached via induction. That is, how can we logically justify predictions we make about events in the future based solely upon regularities inducted from the past? As Hume put it:

> The contrary of every matter of fact is still possible, because it can never imply a contradiction and is conceived by the mind with the same facility and distinctness as if ever so comformable to reality. *That the sun will not rise tomorrow* is no less intelligible a proposition and implies no more contradiction than the affirmation *that it will rise*. We would in vain, therefore, attempt to demonstrate its falsehood.

If we take refuge in the argument "well, predictions I have made in the past based upon induction have proved more often right than wrong," we have done nothing more than rely upon induction as a justification for itself, and thereby we commit the fallacy mentioned by Northrop for deductive formulations. We end up affirming the consequent.

We do not pretend to present an exhaustive treatment of the logical problems involved with the doing of science and the construction of scientific theories, or predictions based upon empirical generalizations. What we do wish to offer is an explanation of how these problems arose. To do so we must refer to our discussion of the empirical modification cycle. Prior to the cognitive extension of prehension, our prehominid ancestors formulated through prehension deductive-inductive models in their brains. The net effect of this process was simply efficient adaptation and survival (*à la* the Scriven theorem). Models derived initially from the deductive pole of the empirical modification cycle and too widely divergent from external reality were "disconfirmed" by the simple expedient of selection; an erroneous prediction often meant death. A "confirmation" of a model meant continued survival. A successful species was and is one with the capacity to formulate models of reality that allow for correct predictions time after time and generation after generation. The same may be said for predictions based upon models derived initially from the inductive end of the cycle. Predictions based upon past regularities were "confirmed" by continued survival of the species and "rejected" by extinction of a portion or all of the species.

With the advent of the CEP, our hominid ancestors became increasingly capable of arriving at models that encompassed a broad spatio-temporal span. And the greater the span became, the further removed predictions based upon the models became from primordial "confirmation"—that is, from the direct action of natural selection. To put it another way, transitional man began to formulate models of the real world with little or no adaptive significance at all. Thus the confirmation of his models took on the appearance of a tautological feedback loop with no mediating interjection from natural biological selection. The situation, we recognize, becomes ironical. Modern man inherits a biologically structured capacity for inquiry and model building. Yet for most of the models he consciously constructs, he has no biological "red pencil" to tell him when they are correct representations of reality and when they are not. As a result, he formulates extremely variant explanations for the same phenomenon, such as death due to "sprites" and "viruses."

This is why scientists have so strongly emphasized positivism and the attendant *role of prediction* in the process. The scientific revolution really amounts to the reestablishment of some set of logical or normative analogues to replace natural selection. Men of science, whether aware of the process or not (and generally they are not), have been forced to the position of positivism, which states that the value of any scientific model may be measured by the value of the predictions and/or retrodictions it generates. In a sense, a scientific model that generates accurate predictions "survives" as long as it does so about problems deemed relevant by the scientific community involved, and it reaches its demise when a positivistically better model supersedes it, or when the problems considered relevant by the scientific community are changed beyond the capacity of the model to solve.

Subsequent to the CEP, and subsequent to reaching optimal adaptation with the new cognitive structure, members of human societies began to diverge in the level of cognitive complexity of which they were individually capable. This divergence continued, probably accelerating after the Neolithic, until today there exists a vast range of cognitive complexity among members of any particular society, and probably in the average complexity as compared between societies. Levels of cognitive complexity, it should be emphasized, are real and measurable (see Harvey, Hunt, and Schroder, 1961; Schroder, Driver, and Streufert, 1967; and Schroder, 1971). Furthermore, variations in levels of cognitive complexity, both within a society and between societies, are due to the genetically endowed capacity to modify models, on the one hand, and more importantly, the richness of experience interactive with models during ontogenesis. As we have stated a number of times, it is not our intent to discuss ontogeny of cognitive systems or models in this book. Indeed, it is a complex and, at times, controversial topic. However, the reader must be aware that there is a degree of plasticity in the human central nervous system, especially during its early ontogenetic phases, and that this plasticity emerged, ultimately, as a by-product of effective, post-CEP adaptation.

References

Driver, Harold E. 1966. "Geographic-Historical vs. Psycho-Functional Explanations." *Current Anthropology*, 7(1).

Feyerabend, Paul. 1965. "Reply to Criticism." In *Boston Studies in the Philosophy of Science,* (ed. by R.S. Cohen and M.W. Wartofsky), Vol. 2. New York: Humanities Press.

Goodenough, W.H. 1963. *Cooperation in Change.* New York: Russell Sage Foundation.

Harvey, O.J., D.E. Hunt, and H.M. Schroder. 1961. *Conceptual Systems and Personality Organization.* New York: John Wiley.

Hempel, Carl G. 1966. *Philosophy of Natural Science.* Englewood: Prentice-Hall.

Homans, George C., and David M. Schneider. 1955. *Marriage, Authority, and Final Causes: A Study of Unilateral Cross-Cousin Marriage.* Glencoe, Ill.: Free Press.

Horton, Robin. 1967. "African Traditional Thought and Western Science." *Africa,* 37(1):50-71.

Hume, David. 1955. *An Inquiry Concerning Human Understanding.* New York: The Liberal Arts Press.

Kaplan, Abraham. 1964. *The Conduct of Inquiry.* Scranton: Chandler Publishing Company.

Korzybski, Alfred. 1958. *Science and Sanity.* (4th edition), Lakeville: International Non-Aristotelian Library.

Kuhn, Thomas S. 1970. *The Structure of Scientific Revolutions.* Chicago: University of Chicago Press.

Levi-Strauss, Claude. 1966. *The Savage Mind.* Chicago: University of Chicago Press.

——1967. *Structural Anthropology.* Garden City: Doubleday.

——1969. *The Elementary Structures of Kinship.* Boston: Beacon Press.

Needham, Rodney. 1962. *Structure and Sentiment.* Chicago: University of Chicago Press.

Northrop, F.S.C. 1947. *The Logic of the Sciences and the Humanities.* New York: Macmillan.

Piaget, J. 1973. *The Child and Reality.* New York: Grossman.

Radcliffe-Brown, A.R. 1924. "The Mother's Brother in South Africa." *South African Journal of Science,* 21:542-555.

Radin, Paul. 1957. *Primitive Man as Philosopher.* New York: Dover Publications.

Schroder, H.M. 1971. "Conceptual Complexity and Personality Organization." in *Personality Theory and Information Processing* (ed. by H.M. Schroder and P. Suedfeld). New York: Ronald.

Schroder, H.M., M.J. Driver, and S. Streufert. 1967. *Human Information Processing.* New York: Holt, Rinehart and Winston.

Service, E.R. 1962. *Primitive Social Organization.* New York: Random House.

Silverman, I. 1964. "In Defense of Dissonance Theory." *Psychological Bulletin,* 62:205-290.

Skyrms, Brian. 1966. *Choice and Chance.* Belmont: Dickenson Publishing Co.

Whitehead, A.N. 1957. *Process and Reality.* New York: Harper.

——1964. *Concept of Nature.* Cambridge: Cambridge University Press.

VII

Structuralism and Language Acquisition

> It is true that a discipline whose main, if not sole, aim is to analyze and interpret differences evades all problems when it takes into account only similarities. But at the same time, it thus loses the means of distinguishing between the general truths to which it aspires and the trivialities with which it must be satisfied.—LEVI-STRAUSS, 1967

IN THE PRECEDING CHAPTERS we have put forth the strong position that all men everywhere process sensory input in more or less the same fashion. Neurognostic models are inherited through the recombination of genetic material in the gametes and are later modified and ramified in reaction to feedback from the external world. Other models are developed in memory solely as a gnostic response to external feedback. These models are being continually tested for accuracy of fit by formation of probability expectations—expectations that are homologous to deductive hypotheses and that are matched to empirically perceived outcomes. The result is usually either the authentication of the holographic models or neurophysiological refinement of the models so that future expectations more closely approximate predicted events and relationships.

We have also argued that men are conscious of only a minor fraction of this process and, as a result, have derived erroneous and often preposterous notions of how they function as neurophysiological phenomena. Thus we have embedded in our philosophical tradition a vast set of epiphenomenal concepts that make it difficult for science to relate observations of "social fact" with internal brain states; e.g,, rational-irrational, sane-insane, intuition-reason, and a bevy of other oppositions, probably the most misleading of which, from the present point of view, are the nature-nurture and mind-brain dichotomies.

The Nature of Structures

The great contribution of Levi-Strauss (1966) has been to show the structural equivalence, at the level of "primitive thought," of a myriad of sociological phenomena for which we have different terms and that we thus consider as truly different things. An example of the application of his insight is the derivation of the functional equivalence of systems of totemic prohibitions and prescriptions among certain Australian societies on the one hand and caste systems of India on the other (1966, chapter 4). Confused both by the limitations of our awareness and by the complexity of "cultural" elaborations having little or no adaptive value, we have found it difficult to center on the underlying and essential similarities among human groups. Structuralism has offered us a conceptual turning point here, demonstrating, as it does, the isomorphism between structure and reality. Yet we intend to carry the structural model further than previous structuralists have seen fit thus far to go. We argue that the underlying structures are, in fact, genetically based, and, as such, they are the result of ages of selective pressures extending back beyond the point in time when man acquired the use of tools and language.

The reluctance on the part of Levi-Strauss and his followers to cross the boundary from the social to the biological has caused a serious weakness in previous structural models, Levi-Strauss, himself, says:

> I still believe that the prohibition of incest is to be explained entirely in terms of sociological causes. Certainly, however, I have treated the

> genetic aspect in too casual a manner. A more accurate appraisal of the very high rate of mutations and the proportion which is harmful would call for some qualification of my statements, even if the deleterious consequences of consanguineal unions have played no part in the origin or the persistence of rules of exogamy. On the subject of biological causality I shall now do no more than repeat that social anthropology has no need of this hypothesis to explain marriage prohibitions. (1969, p. xxviii)

He would, following as he does in the footsteps of Durkheim, and given his central dichotomy between nature and culture, extend this reluctance to any topic of sociological relevance. Because of an antibiological (antireductionist) bias, Levi-Strauss is forced to resort to a fuzzy idea of rationality in order to explain the origin of structural elements. Structural elements are generated by "regulating principles" located at the structural level: A regulating principle

> . . . can have a rational value without being rationally conceived. It can be expressed in arbitrary formulas without being itself devoid of meaning. It is in the light of such considerations that the relationships between dual organizations and cross-cousin marriage should be defined. We do not simply propose to inverse the hypothesis of the priority of dual organizations over this form of marriage. *We believe that they both have their origin in the apprehension, by primitive thought, of those completely basic structures on which the very existence of culture rests.* In this sense, it can be said, but only be said, that cross-cousin marriage and dual organization correspond to different stages in the growing awareness of these structures, and that the practice of marriage between cross-cousins, which appears more as an attempt than as a codified system, constitutes, from this psychological viewpoint only, a process which does not require so complete and definitive an awareness as the institution of dual organizations. (1969, p. 101; emphasis added)

From the biogenetic structuralist's point of view social institutions are behavioral exemplars of an underlying neurophysiological structure that regulates human social behavior in a biologically adaptive manner. Levi-Strauss' "regulating principle" is both (i) the cognitive structure—that is, the holographic model within the brain that, for example, determines a tendency for persons and entities at the social level to be organized according to certain general patterns —and (ii) the biological function (including the adaptive coefficient) of the structure. Instead of relying upon metaphysically laden concepts like "awareness," "rational value," "apprehension," and "primitive thought," he might have seen that prehominid group members whose brain circuitry allowed for, or compelled, a tendency toward organized intergroup exchange gave their groups survival advantage over groups not developing such circuitry. Levi-Strauss might have, to put it in a word, relied more completely on the mechanism of natural selection for an explanation of first causes.

Yet the problem with attributing initial causation in the work of Levi-Strauss, as well as of others he has influenced, is in the ontological status of his structures. Levi-Strauss is a Neo-Platonist who conceives of structures as "ideal forms" having, in some sense, an existence apart from prehending minds. He is thus able to write in the quotation above ". . . that cross-cousin marriage and dual organization correspond to different stages in the growing awareness of these structures. . . ." Traditional structuralists posit, either explicitly or implicitly, the existence of a realm of ideal structures that men, at various places and at various times, are capable of partially prehending (see Figure 7.1). Various patterns manifest in different social structures will be the behavioral by-products of the degree to which the relevant ideal structure is prehended. The only way social scientists can isolate and analyze the ideal structure behind the apparent one is by picking up on the commonalities in manifest structures and then deriving the ideal structure by deduction.

The biogenetic structural position expressly rejects the Neo-Platonist assumptions underlying traditional structuralism. In the first place, the traditional scheme often (but not inevitably) generates explanatory propositions that are, in principle, untestable. In the second place, deductive formulations presented in explanation of such phenomena as matrilateral cross-cousin marriage, marriage classes, mythical themes, initiation, and totemism, though they may genuinely be treated as sufficient conditions for the observed phenomena, are normally treated as necessary as well. Structural arguments are thus often based upon *post hoc* or teleological reasoning. It has been argued that the explanations of Levi-Strauss and company constitute a "new science," presumably in opposition to positivism (Ardener, 1971; Prattis, 1972). The biogenetic structuralist is

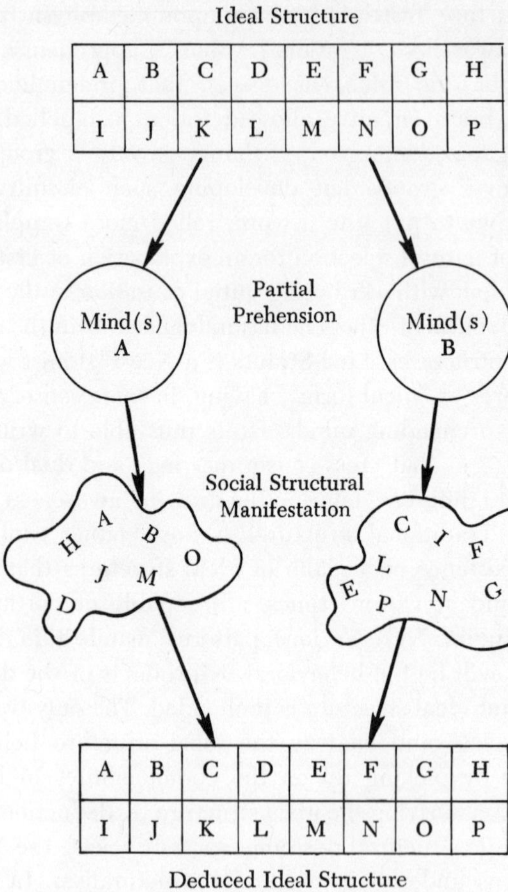

Figure 7.1. Orthodox Structuralism as a Deductive Operation.

forced to reply that a "new science" is a contradition in terms. A "new model of how sciencing actually occurs" (or "reconstructed logic" to use the term of Kaplan, 1964) would be closer to the case and avoid the misplaced concreteness inherent in such a thesis. But then the alternative thesis would be instantly amenable to empirical verification (Laughlin, 1973).

The regularities underlying diversity in human society are indeed structural in nature. But the structures are neither ideal nor imperfectly prehended. Furthermore, they have an ontological status as "real" as the left ventricle or cornea. In short, they are the neurognostic models discussed at length earlier in this book. We will agree that something of the nature of the neurognostic models may be deduced from commonalities in different human societies. As a matter of fact, at the present stage of the development of neurophysiological technology, we often have no recourse but to deduce from behavioral manifestations of internal brain states. Yet we must realize that to argue from behavioral equivalents is to commit an unavoidable tautology. We must assume within the body of our argument precisely the relationship we are attempting to verify. Fortunately not all the evidence for the biogenetic structural position requires the leap from behavior to internal state, as we shall see later in this section and in the following chapter.

Social and Linguistic Universals

It is widely believed that variation in human social organization is due wholly to the patterns of learned behavior generated in and by each society. This is in contrast to the presumed lack of variation in the social structure of prehominids. The latter conclusion is based, in part, upon inductive inference from traditional primatological studies. Work during the past decade in primatology, however, has shown primate social structure to exhibit a much greater variation than was supposed, and this variation is partially in response to the vicissitudes of environment, principally food supply and prevalence of predation (see Denham, 1971). Itani and Suzuki (1967), reviewing the evidence of field observations for the chimpanzee, point to the remarkable plasticity in size and organization of chimp groups. They conclude that this variety of social organization probably offered an optimal adaptation to local conditions in the savanna environment where it likely developed. If they are correct, this offers a striking parallel to hominid evolution, for it is probable that a range of variation in social organization also characterized the prehominid group as a potent adaptive response to savanna living, a point also made by Kortlandt and Kooij (1963).

With the inception of the cognitive extension of prehension and cross-modal transfer, the range of variation in social organization would have become expanded and each variant more complex owing to the increased capacity of hominids to predict environmental shifts, predictions based upon an inductive temporal prehension extended beyond common sense. Organizational patterns emerged based upon the capacity to predict cyclical shifts in basic resource availability. It makes no sense from the present perspective, however, to talk of increased variation as some sort of alternative to inherent behavior. Man did not suddenly drop his set of genetically derived models for a more adaptive set based solely upon "learning." Human social reality can no longer be explained by use of such simple-minded dichotomies. With the cognitive extension of prehension, the holographic models became ramified and more complicated, but they did not lose their genetic base, i.e., neurognosis and neurognostic structure. The genetic basis of human social behavior only became more obscure.

The existence of social universals among human groups has long been an uncomfortable irritant for social and cultural anthropological theorists. We are offered only two general competing explanations for these phenomena: They are either due (i) to historical diffusion or (ii) are considered to be "functionally" necessary for the existence of society. Both explanations are usually *post hoc* and/or tautological in application, and neither is overwhelmingly compelling in this modern age. Katz (1972) has considered the historical development of the sociocultural universals in considerable detail. The reader is referred to his "Unity and Diversity of Man from the Point of View of Social and Cultural Anthropology" for a thorough treatment of this particular issue. We are forced to examine a third and more radical explanation, that offered by the biogenetic structural approach. The approach has important methodological ramifications, not the least of which is that it necessitates a turnabout from the traditional concern with the exotic to a serious examination of the similarities among societies. Similarities will, as often as not, be obscure, but their discovery is important. Similar patterns offer critical clues to the nature of neural structure and cognition. Specifically, this theory predicts that most social (and linguistic) universals are manifestations of genetically determined neurognostic models, or sets of such models, in the deep memory structure of the human brain.

Most of the research work relevant to this line of reasoning has been in the area of linguistic universals (see Greenberg, 1966), in universal semantic fields and in universal patterns in the acquisition of grammar. The findings of Berlin and Kay (1969) are especially noteworthy. In their research into the nature of color categories, Berlin and Kay discovered a maximum of eleven basic color terms present in any of the sample languages examined. More importantly, and as an unpredicted consequence of their study, they discovered that these eleven color categories scaled. They found that if a language has two color terms, they will be white and black. If the language has three terms, they will be white, black, and red. If four terms, the previous three plus either green or yellow, and so on (p. 255). Clearly some basic proclivity exists for the brain to classify chromatic input in a particular hierarchial fashion, and it is this hierarchy that is reflected in the evolution of classificatory systems. Berlin and Kay cautiously suggest that the push for increased complexity of linguistic expression of this hierarchy comes from increased social complexity and the concomitant necessity for more discriminating linguistic categories (p. 16). The critical point here is that as the language changes in response to increased social complexity (or some other social variable), it does not do so at random; nor does it depend upon historical vicissitudes, but reflects and exposes the underlying neurognostic models related to chromatic input. It is critical to understand that Berlin and Kay's results should not be considered atypical of the nature of human language. It is true that Berlin and Kay were surprised by some of their findings (p. 2). Far from surprising, however, we suggest that their results will be repeated time and again in the future as we seek out the underlying structure of language.

The theory of biogenetic structuralism would predict that all languages are grounded in the same, biogenetically structured, holographic models and that this fact would, for one thing, explain the relative ease with which one language may be translated into another. Traditional linguists, in support of the cultural relativism of language, have often pointed to such words as the German *Schadenfreude* (the most succinct definition of which in English is "one who takes malicious joy in another's misfortune") as examples of the difficulties confronted by a translator in finding equivalent glosses. The point missed is that the type of literature usually being translated is rife with complex and ethnocentric allusions. In other words, it is the word *Schadenfreude* that causes the translator a problem, not the words *Himmel* ("sky"), *Mund* ("mouth"), *laufen* ("to run"), and *toten* ("to kill"), or in fact most of the words with which he is confronted. What we wish to show by this is that holographic models would function to provide something equivalent to Chomsky's "deep structures" (1965; 1968, p. 6). The neurognostic portions of these models would be comprised of a complex of circuitry capable of the myriad of associative "transformations" necessary for generating a seemingly endless variety of lexical forms at the behavioral level. Yet a core of semantic forms with a high probability of occurrence in any language is likewise provided.

Another important area of research that promises to yield data in support of the present theory is that of language acquisition. The relevance of this research is already indicated by McNeill's (1970. pp. 52 ff.) discussion of the evolution of basic grammar. Following an intriguing notion suggested to him by G. A. Miller, McNeill argues that hominid reliance upon verbal communication is directly linked to evolution of the brain. As the brain grew larger, and as the pelvic opening grew smaller due to changes necessitated by the shift toward bipedalism, a birth earlier in the maturation cycle resulted, producing, in turn, increasingly dependent offspring. Newborns, because they lack sufficient motor control to signal their needs to their mothers through kinesic means, were thus required to signal by means of their "exclamatory system" (1970 p. 53). The selective pressures would have been enormous for an elaboration of this form of communication, a selection process that may explain the puzzle mentioned in an earlier chapter about the development of the speech center in the brain.

Indirect evidence tending to corroborate both our position and McNeill's has been developed by researchers at the Karolinska Institute in Stockholm, Sweden. Examining the nature of infant cries, a team of medical scientists have discovered that particular cries emitted by diseased and brain-damaged infants give reliable indications of type of abnormality. So far, brain damage, as well as the "Cri du Chat" and Down's syndromes (O. Wasz-Hockert et al., 1968), and hyperbilirubinemia (O. Wasz-Hockert et al., 1971) have been fully analyzed. O. Wasz-Hockert and his colleagues (1971, pp. 4 ff.) also found that cries of normal neonatal infants readily indicate such internal physioemotional states as pain, hunger, and pleasure. The evidence shows, in rough measure, that neonatal baby cries are anything but random noise—that they are, in fact, (i) semantically structured and (ii) behavioral manifestations of genetically based neurognosis. We would predict that infant cries prove to be even more semantically structured than findings show at present. It is likely that, as man developed complex speech, he lost at least in part, the ability to "read" the language of infants. This would tend to explain the relative unintelligibility of a baby's cries to many adults.

McNeill (1970 pp. 70 ff.) also offers a discussion of Chomsky's "Language Acquisition Device" (LAD), which is the hypothetical mechanism in children that stands in the middle of the transition from massive and unintelligible input of communication ("corpus") to the ability to utilize language in communication ("grammatical competence"). He concludes, rightly, that the LAD "is, of course, a fiction" (p. 71) and further states, wrongly, that:

> The question of whether thought affects language cannot therefore be answered in general or in advance. A vacuum exists precisely where speculation, which abhors a vacuum, would most like to enter into the explanation of children's linguistic abilities. (p. 75)

McNeill does go so far as to attribute the basic capacity to acquire language to genetic inheritance (p. 60). Unfortunately, however, McNeill (and Chomsky as well) finds an annoying void precisely for the same reason that Levi-Strauss finds it difficult to talk about the ultimate cause of the incest taboo; they are reluctant to progress beyond the level of linguistic (social) fact to the level of brain circuitry.

The problem of language acquisition is solved, or at least

becomes more intelligible, by substituting vacuum-filling fictions like the LAD, which perpetuate the "black box" status of the brain, with the basic model generating and testing capacity advocated here. Language is prehended by a child in the same fashion as any other auditory input. It is massive and continuous and must be ordered (*à la* the Scriven theorem) so that (i) some paradigmatic basis for distinguishing novel input may be developed, (ii) noise may be filtered out, and (iii) such input can be selectively and adaptively responded to. The circuitry is not selected at random, any more than the circuitry for any of the other holographic models is selected at random. It is the product of untold generations of natural selection. Studies examining maturation of brain morphology and rate of language acquisition have shown a close and almost invariable relationship between the sequencing of both (see Lenneberg, 1966). We would again note the findings of the Karolinska Institute pertaining to semanticity of baby cries and would predict that further research in this area will show revelant, nonrandom correlations between the phonetics and grammar of semantically loaded baby cries and the phonetics, as well as grammar, of adult language.

Progressing from these suggestive genetic predilections, the infant makes his early attempts to speak. These attempts may be seen as behavior based upon probability expectations generated from a maturing set of models within the child's brain, already a mix of genetically given and socially conditioned. The probability expectations are then evaluated in relation to responsive input, and progressive development of his dependence upon linguistic communication are, in a word, tests of holographic models for the child, and responses act as input upon which the models are modified. This is an oversimplification, of course, as indicated by Figure 7.2. Linguistically oriented holographic models may actually be modified by nonlinguistic as well as linguistic input. Thus, at the level of holographic models, linguistic and nonlinguistic structures are virtually equivalent.

The closest theory to biogenetic structuralism in the area of language acquisition is that of Eric Lenneberg. Lenneberg (1967) has built a cogent argument for the existence of a universal grammar dependent upon common modes of biogenetically based cognition.

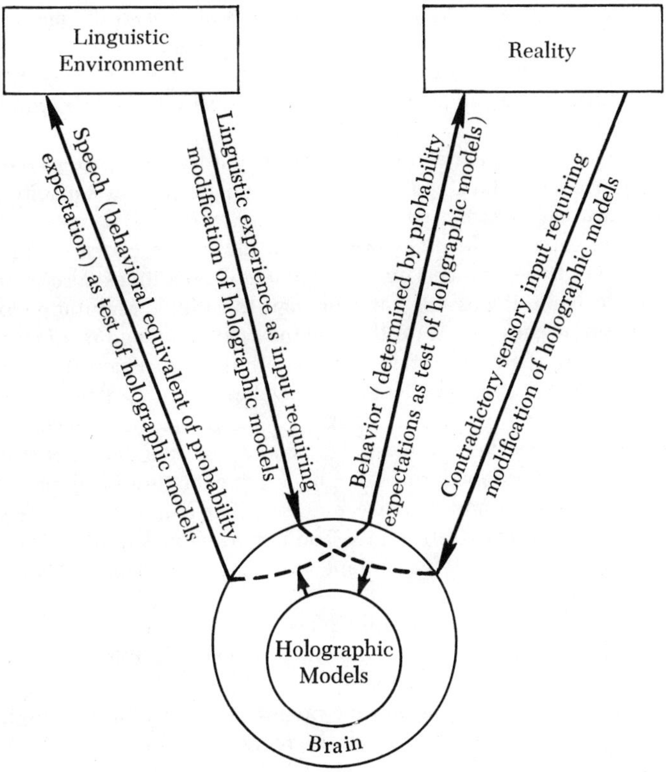

Figure 7.2. The Interaction of Language, Cognition, and Reality.

Because of the proximity to the present theory, it would be well to look at his formulation in detail. It may be presented by quoting partially Lenneberg's thirteen propositions published in *Biological Foundations of Language* (1967, pp. 374–379):

(1) Language is the manifestation of species-specific cognitive propensities.

(2) The cognitive function underlying language consists of an adaptation of a ubiquitous process (among vertebrates) of categorization and extraction of similarities.

(3) Certain specializations in peripheral anatomy and physiology account for some of the universal features of natural language, but the

description of these human peculiarities does not constitute an explanation for the phylogenetic development of language.

(4) The biological properties of the human form of cognition set strict limits to the range of possibilities for variations in natural languages.

(5) The implication of (1) and (2) is that the existence of our cognitive processes entails a potential for language. It is a capacity for a communication system that must necessarily be of one specific type. This basic capacity develops ontogenetically in the course of physical maturation; however, certain environmental conditions also must be present to make it possible for language to unfold. Maturation brings cognitive processes to a state that we may call *language-readiness.* The organism now requires certain raw materials from which it can shape building blocks for his own language development. . . . The presence of raw material seems to function like a releaser for the developmental language synthesizing process. The course of language-unfolding is quite strictly prescribed through the unique maturational path traversed by cognition, and thus we may say that language-readiness is a state of *latent language* structure. The unfolding of language is a process of *actualization* in which latent structure is transformed into *realized structure.*

(6) . . . Actualization may take place even if responses are peripherally blocked; in this case actualization is demonstrable only through signs of understanding language.

(7) The maturation of cognitive processes comes about through progressive differentiation. . . .It might be more fruitful to think of maturation, including growth and the development of behavior such as language, as the traversing of highly unstable states; the disequilibrium of one leads to rearrangements, and so on until relative stability, known as *maturity*, is reached.

(8) The disequilibrium state called language-readiness is of limited duration. It begins around two and declines with cerebral maturation in the early teens.

(9) The language potential and the *latent structure* may be assumed to be replicated in every healthy human being because they are a consequence of human-specific cognitive processes and a human-specific course of maturation. In other words, universal grammar is of a unique type, common to all men, and it is entirely the by-product of peculiar modes of cognition based upon the biological constitution of the individual.

(10) Because latent structure is replicated in every child and because all languages must have an inner form of identical type (though an infinity of variations is possible), every child may learn any language with equal ease.

(11) The raw material from which the individual synthesizes build-

ing blocks for his own language development cannot be the cause of the developing structure as evidenced by the autochthonous beginnings in the infant's language acquisition.

(12) Social settings may be required as a trigger that sets off a reaction. Perhaps a better metaphor still is the concept of resonance. In a given state of maturation, exposure to adult language behavior has an excitory effect upon the actualization process much the way a certain frequency may have an excitatory effect upon a specific resonator; the object begins to vibrate in the presence of the sound.

(13) Even though biological constitution of the individual is an essential replica of its progenitors, there are, naturally, individual variations.

We view Lenneberg's theory as essentially parallel to our own. If we translate his theory into the present terminology, he is positing an empirical modification cycle between potential associative channels emerging in the brain as that organ matures, on the one hand, and the linguistic environment, on the other. "Language-readiness" occurs with the maturation of neurognostic linguistic models ("latent structures"), probably primarily located in the Wernicke-Broca areal complex and the dominant inferior parietal lobule. After the emergence of the neurognostic models, a period of time remains during which they may be optimally modified by the addition and re-addition of dendritic-synaptic material. After the period of optimal potential growth, though models continue to be modified, they are beyond the capacity for optimal change.

There is an increasing body of evidence from ethology that strongly suggests that this process is a general one, a process that may be characteristic of species that specialize in vocal communication. A number of authorities (Hinde, 1969; Marler, 1970; Nottebohm, 1972; Smith, personal communication) have noted that many bird species exhibit a "critical period" (or "sensitive period") during which fully developed song must be learned if it is to be learned at all. The length of the critical period varies according to species (Hinde, 1969, p. 25), but the overall phenomenon across species is fairly homologous.

A young bird must experience the song of its parents or conspecifics during the critical period in order to develop normal mature song. If it is experimentally isolated from such experience during the critical period, a bird will make vocalizations at maturity, but the song patterns it reproduces are rough and bear little relation to mature adult song. More than this, the "number and time of exposures to the song" are important to proper development (Marler, 1970, p. 16). Finally, if a bird is experimentally deprived of exposure to adult song for the entire critical period, it will never learn mature song, even if the bird is later exposed to song under normal conditions (Marler, 1970, p. 17).

These findings, in combination with the fact noted earlier (chapter 5) that birds evidence a degree of preparedness in learning some vocalizations, suggest the clear participation of neurognosis and the empirical modification cycle. As discussed by Hinde (1969, p. 25), the problem for birds seems to be to match vocal motor skills as young adults to models they obtained somatosensorally as chicks. The latter models are derived *in toto* previous to any attempt to behave in accordance with them. We would hypothesize that for many birds the acquisition of normal adult song requires an ontogenic modification of programmed neurognosis through the empirical modification cycle. This first phase is probably mediated by the gnostic cortex. During a later phase, models in the motor region of the bird's brain, specifically those nuclei innervating and controlling organs of vocalization, are in turn modified through the empirical modification cycle by matching vocalizations (as probability expectations) against the gnostic models. It has been noted by W. J. Smith (personal communication) that a bird's first attempts at adult song are imperfectly patterned, but that the bird quickly develops greater facility through use.

Although it does not relate directly to the ontogeny of language acquisition, it might be well at this point to mention another development in the ethology of communication that promises to have an impact on our understanding of communication in man and other animals. W. J. Smith and his associates have demonstrated for three species of birds[1] that song often provides information by various permutations of a finite number of discrete calls. That is, song for birds has the quality of a *string*; the information is provided, not by

[1] Eastern Phoebe or *Sayornis phoebe* (Smith, 1969a); Eastern Wood Pewee or *Contopus virens* (Smith, personal communication); and Yellow-Throated Vireo or *Vireo flavifrons* (Smith, in preparation).

the discrete calls themselves, but by the proportion of one call in relation to another call. Thus, if we hypothetically isolate one call and label it "1" and another call and label it "2", the two sequences of calls represented in the following schematized songs

211211211211211 2
211112111121111 2

would provide two distinct messages. It seems likely to us that this principle will prove to be a general one for animals specializing in vocal communication. At some point in the development of the specialization an animal will begin to rely upon concatenations of a finite number of distinct vocal units, rather than continue to add distinct calls *ad infinitum*. The selection pressure would probably be in favor of greater facilitation of redundency (thus less likelihood of misunderstanding on the part of the receiver) and decrease in "noise" (thus greater precision of information transmission). Indeed, Smith (1969b:149) has discovered that the number of discrete vocal and other displays available to any species for which there is information varies from 15 to only around 45. These findings when considered in light of the limited number of phonemes utilized in any human language are both striking and suggestive.

Predictions From Biogenetic Structural Theory

It is clear that the theory propounded here acts to tie together a number of apparently disparate theoretical positions, certainly more than specifically discussed here. In particular, it represents a synthesis of traditional structural and evolutionary theory. A large number of specific predictive hypotheses are deducible from the model. We have chosen, however, to center on only a few possible areas of concern in the interests of space. They will occupy the rest of this chapter and the whole of the next.

For some reason, anthropologists have totally missed the relevance of a study of sleep to anthropological theory. Perhaps this has been overlooked for the same reason that other universals have been missed, because anthropologists have been busy looking for the non-universal and exotic. The first relevant fact about sleep is that it is, as we all know, universal. All men (all higher organisms as well) must sleep. Hence all societies must be structured so that all members obtain sleep. Exactly when men sleep, where they sleep, upon what they sleep, what ritual sanctions surround sleep, etc., do not matter here. These are merely elaborations on the basic structural theme. What matters is that men sleep, and the relevant question is *why men must sleep*?

We have long known that humans, and other species, do not need sleep merely to recuperate from physical fatigue. This function requires but little time relative to the amount of sleep required by the organism. Man needs sleep in order to deal in some way with sensory input during waking hours (Luce and Segal, 1966, chapter 1). Early studies on persons systematically deprived of REM sleep indicated that the REM phase was necessary in the maintenance of sanity (1966, pp. 85 ff.).[2] REM-deprived individuals exhibited temporary symptoms of psychosis. More recent studies seem to associate less serious abnormal behavior with REM deprivation (Van de Castle, 1971, pp. 30, 31). Research findings, at the present time, seem to be confused on the issue. Furthermore, a great deal of controversy is centering around the theoretical implications of the interpretation of dreams. We wish to avoid a discussion of most of these weighty issues at this time. However, the model developed here offers a partial explanation of the relationship between model construction and sleep. We will briefly describe this relationship and offer some testable hypotheses relating to sleep.

Contradictory input into the brain through prehension requires the modification of models within the brain relating to that input. That is, input contradicting expectations will cause proportional physical modification of the dendritic-synaptic structure comprising the model. Rapid or massive contradictory input will require massive modification of the internal models. Slight input will require that the models be modified only slightly. We would predict that

[2] REM ("rapid eye movement") sleep refers to that part of sleep when eye movements and particular EEG readings presumably indicate one state of dreaming.

the modification of dendritic-synaptic structure, because it is a process of physical growth of neural cells, requires a period of release from sensory input in order to "catch up" with previous input before again generating expectations and absorbing new input. In other words, an outside limit exists to the rapidity with which the models may be physically altered. The predictive ramifications of this are numerous:

(i) We would expect to find at least two levels of "memory," one that contains the holographic models *per se* and another that holds contradictory input requiring model modification in storage until the first level is able to deal with it.[3]

(ii) We would expect to find a close correlation between the amount of input contradictory to expectations generated by the models and the amount of REM sleep required by the subject.

(iii) We would expect gradual "withdrawal from reality," or a severing of the input-to-model–model-to-expectation cycle in subjects systematically deprived of REM sleep. This prediction is reached by reasoning that if the modification of holographic models to match contradictory input is allowed to fall too far behind the flow of input from prehension, there will develop a serious mismatch between the models and reality, a rapid increase in the inadequacy of generated expectations to predict reality, and hence a marked increase in contradictory input. If left unchecked by the brain, the models would become totally unrepresentative of reality and input would become virtually random *vis à vis* novelty/redundancy.

(iv) We would expect that the amount of REM sleep in absolute time required by human subjects would, on the average, decrease with age from birth to death. The models inherited by the newborn are inadequate in their representation and prediction of the input with which they are confronted. Thus the infant will require more REM sleep for the growth of its models than will an old person, say, who finds most input redundant.

(v) We would expect to find a decreasing REM sleep per unit of time differential between newborn and adult in species of decreasing neurophysiological complexity. The more complex the

[3] Short-term storage probably is facilitated by biochemical changes at the synapse (see Deutsch, 1971).

capacity to form holographic models of reality, the more time will be required to ramify the dendritic-synaptic structure and the greater will be the differential in adequacy of models between the newborn and the adult. Thus the differential between newborn humans would be greater than between the two categories among chimpanzees. The differential in chimps would be greater than in baboons, in baboons greater than in lemurs, in lemurs greater than in frogs, etc.

(vi) Finally, we expect to find a relationship between intelligence and duration of REM sleep. We have already defined "intelligence" as the capacity for novel decisioning. Viewed from a slightly different perspective, intelligence may be seen as the capacity to integrate contradictory input with probability expectations through physical modification of the relevant models. Since it seems probable that such modification occurs during REM sleep, the connection between REM sleep and intelligence becomes crucial. A test of the relationship between the two would require a two-step program of experimentation during which intelligence is measured by the Merrill-Palmer test and the rate of external sensory input is controlled:

> Step 1. Null Hypothesis: No significant relationship exists between intelligence and duration of REM sleep.

Step 1 H_0 is assumed to be rejected, experimentation should proceed to test:

> Step 2. Null Hypothesis: No significant relationship exists between high intelligence and duration of REM sleep.

If Step 2 H_0 is assumed to be rejected as well, the experimental results would indicate one of two logical relationships: either high-intelligence subjects exhibit longer durations of REM sleep or they exhibit shorter durations of REM sleep measured against average-intelligence subjects. We suspect that, according to data already available on the nature and function of REM sleep, there will prove to be an inverse correlation between the requirements of high intelligence and minimal REM sleep. The suggested, expected results are depicted in Figure 7.3.

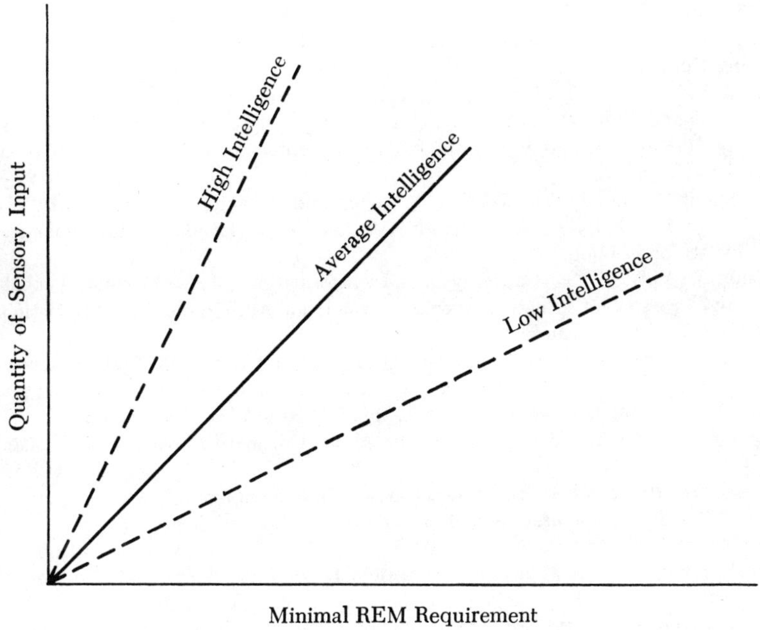

Figure 7.3. Expected Relationship Between Rate of Sensory Input, Intelligence, and Minimal REM Requirement.

References

Ardener, E. 1971. "The New Anthropology and Its Critics." *Man* 6(3):449-467.

Berlin, Brent, and Paul Kay. 1969. *Basic Color Terms: Their Universality and Evolution*. Berkeley: University of California Press.

Chomsky, Noam A. 1965. *Aspects of the Theory of Syntax*. Cambridge: M.I.T. Press.

Chomsky, Noam A., and Morris Halle. 1968. *The Sound Pattern of English*. New York: Harper and Row.

Denham, Woodrow W. 1971. "Energy Relations and Some Basic Properties of Primate Social Organization." *American Anthropologist*, 73(1):77-95.

Deutsch, J. Anthony. 1971. "The Cholinergic Synapse and the Site of Memory." *Science*, 173:788-794.

Greenberg, Joseph H. 1966: *Language Universals*. The Hague: Mouton.

Hinde, R.A. 1969. *Bird Vocalizations*. Cambridge: Cambridge University Press.

Itani, J., and A. Suzuki. 1967. "The Social Unit of Chimpanzees." *Primates,* 8:355-381.

Kaplan, Abraham. 1964. *The Conduct of Inquiry.* Scranton: Chandler Publishing Company.

Katz, S.H. 1972. "The Unity and Diversity of Man from the Point of View of Social and Cultural Anthropology." Presented at seminar on Unity and Diversity of Man, Fondation Royaumont, Paris, September. Also in *The Unity of Man* (in press).

Kortlandt, A., and M. Kooij. 1963. "Protohominid Behavior in Primates." In *The Primates* (ed. by J. Napier and N.A. Barnicot). London: The Zoological Society of London.

Laughlin, C.D. 1973. "Causality and Transformational Law: A False Dichotomy." Presented at annual meeting, American Anthropological Association, New Orleans, November.

Lenneberg, Eric H. 1966. "The Natural History of Language." In *The Genesis of Language* (ed. by F. Smith and G.A. Miller). Cambridge: M.I.T. Press.

——1967. *Biological Foundations of Language.* New York: John Wiley.

Levi-Strauss, Claude. 1966. *The Savage Mind.* Chicago: University of Chicago Press.

——1967. *Structural Anthropology.* Garden City: Doubleday.

——1969. *The Elementary Structures of Kinship.* Boston: Beacon Press.

Luce, Gay Gaer, and Julius Segal. 1966. *Sleep.* New York: Coward-McCann.

Marler, P. 1970. "A Comparative Approach to Vocal Learning." *Journal of Comparative and Physiological Psychology,* 71 (2, Part 2) 1-25.

McNeill, David. 1970. *The Acquisition of Language.* New York: Harper and Row.

Nottebohm, F. 1972. "Neural Lateralization of Vocal Control in a Passerine Bird II. Subsong, Calls, and a Theory of Vocal Learning." *Journal of Experimental Zoology,* 179:35-50.

Prattis, J.I. 1972. "Science, Ideology, and False Demons: A Commentary on Levi-Strauss Critiques." *American Anthropologist,* 74(5):1323-1325.

Smith, W.J. 1969a. "Displays of *Sayornis phoebe.*" *Behavior,* 33(3-4):283-322.

——1969b. "Messages of Vertebrate Communication." *Science,* 165:145-150.

Van de Castle, Robert L. 1971. *The Psychology of Dreaming.* New York: General Learning Press.

Wasz-Hockert, O., et al. 1968. *The Infant Cry: A Spectrographic and Auditory Analysis.* London: Heinemann Medical Books.

——1971. "Spectrographic Analysis of Pain Cry in Hyperbilirubinemia." *Biologia Neonatorum* 17:260-271.

VIII

Psychopathology and Evolutionary Structuralism

Two main categories of mental disease are recognized . . .
(1) the organic brain syndromes, a group of disorders caused by or associated with impairment of brain tissue function;
(2) disorders not attributable to organic conditions and presumably of psychogenic origin or without clearly defined physical cause or structural change in the brain. The respective symptomologies overlap almost completely . . . .—ANTHONY F. C. WALLACE, *Culture and Personality*.

IN THIS CHAPTER we shall consider some of the major forms of psychopathology from the biogenetic structural perspective. We shall first analyze schizophrenia as a generalized defect of the gross structural (neuroanatomic) relationships described in chapter 3. In so doing, we hope that our understanding of the genetics of schizophrenia will shed some light on the interrelated evolution of those cerebral structures. We hope, then, that not only will schizophrenia be examined within an evolutionary perspective, but the evolution of higher cortical adaptation in general may also be conceptualized in terms of a coherent genetic mechanism.

In the second part of this chapter we shall consider phobias, depression, alcoholism, and obsessive-compulsive traits in terms of the highly complex inherited cerebral structures that were considered in chapter 5. Last, we shall devote a few words to the concept of "critical periods" in neural development.

Schizophrenia

We may now turn our attention to schizophrenia, and particularly to what is known about the genetic determinants of this disease, so that we may be able to relate this entity to cerebral structure and higher cortical function. Fully to understand the complexity of the problem a genetic theory of schizophrenia must explain, we must give some consideration to the vast scope of symptomatology that has been included under the heading of schizophrenia. These symptoms include looseness of cognitive associations, psychotic ambivalence, inappropriate affect, flatness of affect, poor or absent modulation of affect, spontaneous and repeated blocking of lines of thought, hallucinations, illusions, subjective feelings of separation from reality manifested by various dissociative states, anhedonia or the inability to relate to the environment, paranoid ideation, condensation of the elements of thought, concrete thinking, inappropriate, bizarre, or symbolic associations, uncontrolled agitation, and/or fear and catatonia.

Although it is true that all these symptoms rarely, if ever, occur in the same patient, it is just as true that they do occur in widely overlapping complexes, so that any one symptom has been documented to occur with any other in some patients who carry the diagnosis of schizophrenia. This has caused such experts as Kaplan (1965) to state that patients with a schizophrenic diagnosis, as a group, do not represent a simple category of mental disorder but rather a constellation or even many constellations. Nevertheless, factor analysis seems to indicate that all the symptoms cited above are, in some way, more, or less related. Now all these symptoms can be reduced in one way or another to the cardinal signs of schizophrenia described by Bleuler (1950 edition).

(1) Autism—involving blurring of subject-object differentiation and often accompanied by hallucinations. The inner and outer worlds become confused.

(2) Difficulty with abstraction, usually involving concreteness of thought.

(3) Looseness of associations as verbally expressed by inappropriate linking of elements of speech.

(4) Disorder of affect control as manifested by:

(a) anhedonia—partial or total inability to correlate external objects and persons with a subjective feeling tone. This can present itself as anything from a flatness of affective response in conversation to a total withdrawal with complete lack of interest in the environment.

(b) inappropriate affect, expressed by the manifestation of emotions not in keeping with verbal productions or with environmental circumstances.

(c) ambivalence, expressed by the rapidly alternating sensation of opposite affective states (e.g., love-hate) generated by a single unchanging stimulus in the external environment. It is often defined as the simultaneous experience of opposite affective states.

It is of considerable importance to note that all schizophrenic patients manifest almost all these symptoms if they are assiduously searched out. What is apparently confusing is that the symptoms present themselves in differing degrees of severity in different patients. Thus, for example, one patient may exhibit an extremely flat affect and yet his associations may be only slightly loose. Another patient's verbal productions may be almost incomprehensible owing to looseness, yet his affective response to differing stimuli may be fairly appropriate. In other words, one or two symptoms may predominate in a given patient while the others may remain relatively insignificant. Furthermore, what symptoms predominate and, in what degree, vary from patient to patient. This variability generates the whole confusing spectrum of schizophrenia.

If we consider these cardinal symptoms of schizophrenia with relation to the gross cerebral structures described in chapter 3, it appears increasingly probable that what we are dealing with is partial malfunction of these systems. We noted that according to Luria, frontal projections to the thalamus as well as to all the sensory association areas appear responsible for the self-other dichotomy as well as for integration of internal and external states. Partial malfunction of this system results in a blurring of the self-other distinction with the production of what is clinically called autism. Second, a partial malfunction of the cross-modal transfer mechanism of the inferior parietal lobule results in concreteness of thought and difficulty with abstraction as well as with use of basic logical grammatical categories. This is a symptom in patients with clearly defined lesions in this area (Luria, 1966, p. 154) as well as in schizophrenia. Likewise, lesions in the prefrontal area, anterior to the motor speech area, produce disorganization of thought. This latter condition occurs, in all probability, if the lesion interrupts the frontal projections to the inferior parietal lobule. It is a deficit in this mechanism that probably results in the looseness of associations found in schizophrenia. With respect to affective control, we know that there are frontal projections to the limbic system, especially to the hippocampus (Bonin and Bailey, 1947). Frontal tumors can produce psychotic ambivalence and inappropriate affect identical to that produced in schizophrenia. One is left with the conclusion that the sequential ordering of affective states and its correlation with the cognitive sequence is interrupted by disruption of frontal-limbic connections.

Last, let us consider the inability to cathect or place emotional investment in the external world. We have noted above that such an ability is essential to the formation of social groups. Patients who exhibit this inability probably suffer from a partial malfunction of the direct connections between their sensory association areas and the limbic system. They are, therefore, unable to invest incoming stimuli from the environment with an emotional charge, and hence are "unmotivated" to relate to the environment.

The crucial point to note here is that whereas tumors or other lesions may produce any single symptom of schizophrenia, depending on what connections are interrupted, schizophrenia itself produces all (or almost all) the symptoms, even if some are only minimally expressed. On the other hand, a large lesion in a specific area usually produces a profound or even total deficit in the system affected, whereas in schizophrenia, usually some, and often a great deal, of the function is preserved. Likewise, the function of the various systems affected by schizophrenia tend to wax and wane, whereas with a lesion destroying a system, the loss of function tends to be more fixed.

The upshot of all this is that schizophrenia may represent a partial deficit of *all* the cerebral adaptive mechanisms described earlier. Consequently; neural connections all over the brain are likely to be affected, with the deficit tending to fluctuate in response to physiologic and/or environmental changes. Furthermore, the degree of deficit of one system proportional to the other systems varies from person to person, although all the systems are to some degree affected. We shall propose a specific model to explain these findings subsequently. At this point, however, it is essential to consider the evidence that symptoms observed in schizophrenia are at least partially genetic in origin and, if so, what genetic mechanism is most probable. If one is satisfied that schizophrenia has a genetic basis, we have gone a long way toward demonstrating that all the gross cerebral adaptive mechanisms that we have considered in chapter 3 are intimately related on a genetic level, and that they most likely are evolved in a mutually interdependent fashion. The genetics of schizophrenia, therefore, provides a possible key to understanding the evolution of multiple cerebral adaptive mechanisms in intimate relationship to each other. Thus we will be able to understand the adaptive function of the neocortex as a unity rather than in terms of an aggregate of systems as we did in chapter 3. Schizophrenia is seen, in this model, as a defect of a unitary cerebral adaptation, with various of the constituent subsystems suffering greater or less malfunction.

THE GENETICS OF SCHIZOPHRENIA

Perhaps the first significant breakthrough in the study of the genetics of schizophrenia came in 1946, when Kallmann published his now classic paper on the subject. In this paper he analyzed 691 schizophrenic twin index families. He found 64 to 69 percent concordance between monozygous twins as opposed to only 11 to 16 percent concordance between dizygous twins. The implications of the study were immediately obvious, and it provoked a flurry of controversy in the literature, a flurry that has not yet abated. Several other twin studies have been made, most notable being those of Fisher, of Kringlen, and of Gottesman (Lancet, 1966). They have all indicated a very significant concordance of schizophrenia between monozygous twins, although none has been able to establish the extremely high concordance found by Kallmann. It is now a concensus that the concordance rate for monozygous twins is somewhere in the order of 30 to 40 percent.

The evidence for the genetic basis for schizophrenia, however, does not rest solely with the twin studies. Prout et al. (1956) examined pairs of siblings of the same sex with less than two years age difference, one of whom was diagnosed as schizophrenic. The mothers were determined not to be schizophrenic. The study showed that the two siblings (the future schizophrenic and the normal) showed marked personality and temperament differences from a very early age. These differences could not be associated with any environmental parameters, including psychological trauma. The future schizophrenics were more dependent and less able to deal with traumatic events.

Studies showing a significant relationship between schizophrenia and certain morphological characteristics likewise lend weight to the genetic theory. The most impressive findings in this area are those of Cowie, Coppen, and Norman (1960), who found that schizophrenics as a group differed markedly from normal subjects both with regard to the biacromial diameter and the anatomical androgyny score (3 × biacromical diameter in centimeters minus 1 × biliac diameter in centimeters). In other words, schizophrenic patients of both sexes were found to have much lower (i.e., more female) androgyny scores and smaller biacromial diameters.

If one grants that the above evidence indicates a genetic component, at least as a predisposing factor, one is still faced with the question of the precise mode by which schizophrenia is transmitted. Kallmann himself in 1953 proposed the hypothesis that the disease is passed on as an autosomal recessive (non-sex-linked) recessive (Kaplan, 1965). Intensive subsequent evaluation of numerous twin studies, however, has indicated that such an hypothesis does not explain the frequency of occurrence of schizophrenia in most of these populations. Hence the recessive gene hypothesis finds few adherents today.

Book, in 1953, and Slater, in 1958, proposed that the usual genetic basis for schizophrenia is a single dominant gene (Sc) with about a 25 percent penetrance. Most of the adverse criticism of this hypothesis centered on the difficulty in understanding how such a gene could be maintained in the population at a frequency greater than can be accounted for by mutation alone, since this gene is responsible for a trait that seems to be significantly maladaptive. Most damaging of all, as Moran (1965) points out, is the considerable evidence that schizophrenics leave fewer children than those not developing the disease.

In response to these objections, Huxley, Mayo, Osmond, and Hoffer proposed the theory of schizophrenia as a genetic polymorphism in 1964. It was Huxley's contention that whenever an apparently unfavorable trait is maintained in the population at a frequency greater than can be accounted for by mutation alone, the most likely explanation is that it represents a polymorphism. Although this is perfectly acceptable theoretically, it placed the burden on Huxley and his associates to propose positively adaptive traits in schizophrenia that could counterbalance the significantly lowered reproductive efficiency to account for the maintenance of the gene in the population. Their inability to do so has resulted in a general discarding of the polymorphic theory of the transmission of schizophrenia.

Gottesmann and Shields point out in their 1967 article on a polygenic theory of schizophrenia that classical segregation ratios are simply not found for the relatives of schizophrenics. One possible way out of this dilemma is to regard schizophrenia as being polygenetically determined. Thus the disorder could be treated like a threshold character whose phenotypic appearance would depend on both the number of genes present and the amount of stress. This model would predict the continual appearance of schizophrenics in the offspring of normal parents, the increased risks of schizophrenia in loaded families, and the slow response to negative selection. Most important from a clinical point of view, a polygenic theory of schizophrenia would explain the protean manifestation of schizophrenia in widely overlapping symptom complexes and in varying degrees of severity. In terms of the polygenic theory, schizophrenia could be thought of as simply a part of the genetic load, the *price paid for conserving genetic diversity*.

Since Gottesman's paper, a number of excellent studies by Heston and Denny (1968), and by Kety et al. (1968), among others, all point in the direction of a polygenic mechanism underlying the transmission of schizophrenia, although recently Heston (1970) has returned to a major gene hypothesis (admitting, however, the essential effect of minor genes). In general, however, the polygenic theory seems to be the consensus of the leading investigators in the field today (Rosenthal, 1968).

Let us now consider the question how a polygenic mechanism can generate the spectrum of partially functioning cerebral adaptive mechanisms we call schizophrenia. We must first reiterate that each of the cerebral adaptive mechanisms described in chapter 3 involves at least one tract of nerves connecting one part of the brain with another part. Let us postulate, for the moment, that all the tracts providing the neuroanatomical basis for the cerebral adaptive mechanisms are composed of several functionally different fiber types, somewhat analogous to the A, B, and C fibers of the peripheral nervous system but without *gross* anatomical differences. Furthermore, to make matters simple, let the function of each depend on a single enzyme, a different one for each fiber type. If there were three fiber types, the function of each would be controlled by a single gene. Again, for simplicity's sake, let us suppose a fiber tract connecting two parts of the brain is equally represented by all three fibers. If one gene is defective, one-third of the function of that tract would be destroyed with two-thirds preserved. Now let us suppose that all the various connections responsible for cerebral adaptation have all three fiber types represented, but in *varying proportions* (the proportion of any fiber type possibly relating to the evolutionary antiquity or recentness of the structure). In such a case, malfunction of gene I causes malfunction of fiber-type I. This would be reflected in a decreased efficiency in *all* the connections, but the effect would be most profound in those tracts with a preponderance of fiber type I.

At this point, it is important to note two things. First of all, the other cerebral adaptive mechanisms (fiber tract) would also be affected, even if only minimally, since we have postulated that all have at least some proportion of fiber type I. Second, function is not totally obliterated even in those tracts that have predominately type I, owing to the presence of some types II and III. Naturally a totally different symptom complex would present itself with malfunction of gene II, and yet a third symptom complex with the malfunction of gene III. In those instances when there might be a malfunction of two genes, the symptoms would be more profound. Yet the cerebral adaptive mechanisms would all function to some degree at least, unless all three genes became defective. Such a model as we have presented here has the advantage of agreeing with the accumulating evidence that the spectrum of schizophrenia is polygenetically transmitted.

We must emphasize that our use of three fiber types in our example is purely arbitrary and advanced solely to add simplicity to the model. A multiple-fiber model could function just as well with four, eight, eleven (or whatever number) fiber types. In fact, the greater the number of fiber types comprising the major brain connections in varying proportions, the greater and more varied is the spectrum of schizophrenias explained without resorting to concepts of penetrance or to combination defects (two or more malfunctioning types). A two-fiber type model is conceivable, but would probably require varying degrees of penetrance to explain the vast spectrum of symptoms in the disorder.

Our model has the important theoretical advantage of emphasizing the genetic interrelationship of multiple cerebral structures and of indicating the redundancy of the mechanisms underlying higher brain function. Thus a system of multiple-fiber types provides a considerable margin of safety, guarding man's essential adaptive mechanisms against harmful mutations. A marked degree of redundancy would, in itself, be a sign of the adaptive importance of the cerebral mechanisms described.

Admittedly such a model is highly theoretical, however well it fits the evidence. The theory hinges on the existence of multiple functional fiber types in all or most of the major tracts connecting parts of the cortex.

Is there any evidence whatsoever that such is the case? In 1966 d'Aquili, Chambers, Liu, and McCouch investigated the effects of anesthetics on spinal reflexes of monkeys after hemisection and transection of the spinal cord, lesions in the pyramid (corticospinal tract), and lesions in areas (6 and 8) of the motor cortex. In addition to determining a differential rate of recovery of facilitatory and inhibitory mechanisms after spinal shock, it was also found that reflexes on the involved side actually became resistant to anesthetics. This was interpreted to mean that sprouting had taken place to fill in some of the gaps left by degeneration of descending terminals on the involved side. But since anesthetics, and in particular barbiturates, are thought to exert their effect by inhibiting the Krebs Cycle (the principal oxygen-based energy source), and since the reflexes that had recovered from spinal shock were resistant to the effect of anesthetics, it was postulated that these resistant reflexes reflected a population of neurones that could easily convert from metabolism based on the Krebs Cycle (i.e., oxygen) to one based on an anaerobic path (nonoxygen-based energy source) or else on the hexose monophosphate shunt (a minor alternate oxygen-based energy source). Particularly in analyzing the effect of lesions of the pyramid, which contain almost completely facilitatory fibers, it was inferred that at least two populations of facilitatory neurones exist on the spinal level, one that is totally dependent on the Krebs Cycle for its metabolic activity and one that can readily use alternate paths. The former group apparently recover only minimally from spinal shock whereas the latter do so readily. Although this constitutes no proof that multiple functional fiber types are responsible for cerebral connections, it is nevertheless evidence compatible with, and suggestive of, such an hypothesis.

Further evidence lending support to this hypothesis comes from studies attempting to coordinate lower order neurological signs (such as might be elicited by a careful neurological evaluation) with psychopathology in general and schizophrenia in particular.

One would expect that if multiple fiber types exist at the cortical level, some reflection of this might be manifested at lower levels of the nervous system. Thus one would expect to be able to observe at least some classic neurological signs in the presence of deficits of higher cortical functioning such as is found in schizophrenia. This is precisely what a number of studies have found. Thus Hertzig and Birch (1966) found that 62 percent of their group of 104 psychiatrically abnormal adolescent girls had at least one indicator of primary neurologic abnormality. They state:

> Such abnormality was significantly more frequent in the patients diagnosed as schizophrenic than in those otherwise classified. Moreover, a significantly greater proportion of the schizophrenic patients were designated as abnormal by two or more of the indicators. When compared with psychiatrically unremarkable populations of normal children younger in age, both subgroups (schizophrenic and nonschizophrenic) of the psychiatric series were found to have indications of neurological dysfunction four to eight times more frequently on clinical neurological examination, ten times more frequently for adventitious movement, and two to three times more frequently for auditory-visual integrative deficiency. Subnormality in intelligence was significantly more frequent only in the schizophrenic group.
>
> The results . . . are interpreted as suggesting that primary neurologic abnormality is a major etiologic factor in the development of schizophrenia in children and adolescents. (p. 398)

Subsequent work by Hertzig and Birch (1968) as well as by other investigators (Goldfarb, 1964, Fish, 1961, and Rochford et al., 1970) tends to confirm these findings.

Although we do not construe these investigations as proof of our model, we offer them as strong support that such a model is in fact operative in schizophrenia and, perhaps, in other psychopathological states as well.

THE EVOLUTION OF GENETIC CONTROL OF CEREBRAL ADAPTATION

If our analysis of schizophrenia as a deficit of multiple polygenically determined cerebral structures is correct, we can conclude that all, or most, of those mechanisms presented in the first part of this chapter are genetically linked. The implications of such a view are far reaching. First of all, we can conclude that an integrated "master plan" of interrelated cerebral adaptive mechanisms took place as a result of neural reorganization probably at the subhominid level. Not all the neural connections present as a result of this reorganization were sufficiently developed to be initially functional. If we may use the chimpanzee as our example once again, perhaps only the frontal-visual connections related to problem solving were sufficiently established to allow spontaneous functioning. Subsequent mutations laid down different fiber types according to the "master plan," each mutation adding fibers to each major connection, more to some than to others. Thus with each mutation yielding a new "layer" of fibers, we get a *differential development* of the various connections (structures). As each connection achieved a certain degree of development (involving multiple-fiber types), it became spontaneously functional in the animal resulting in an alteration of behavior and adaptive capabilities. Progressive development of these systems past a point of optimal efficiency probably resulted in a marked redundancy of cerebral adaptation. This served as a safety mechanism, insuring cerebral adaptation against harmful mutation. Much of the 1000 cubic centimeter increase in cranial capacity since australopithecine times can probably be ascribed to the highly redundant nature of the cerebral structures involved. When one considers that there is as much as 1000 cubic centimeter variability in cranial capacity in *living* populations (Holloway, 1966) and, furthermore, that those individuals with the smallest cranial capacities show no significant intellectual deficit over those with the largest, one is struck by the danger of imputing behavioral or cerebral adaptive capabilities to extinct populations on the basis of cranial capacity alone.

The mechanism we have proposed has the merit of explaining the apparent genetic linkage of the various structural adaptive mechanisms of the brain as well as their differential rates of evolution. As we have suggested in chapter 3, the trait that was selected for was probably tool using. All the cerebral adaptive mechanisms that we have considered in this discussion either generate more efficient tool using or else are probably genetically linked to those that do. Thus evolution of the neocortex, although it involves differential rates of growth of its subsystems, nevertheless appears to have progressed in an integrated manner that was laid down by a major neural reorganization at the prehominid level. Selection for the use and retention of tools probably directed the evolution of this complex toward progressively higher levels of problem solving culminating in the present cortical functioning of *Homo sapiens.*

Having indicated the functional unity of the evolving cerebral cortex, as well as the probable intimate connection between individual cerebral structures at the genetic level, we can cease speaking of multiple adaptive mechanisms and refer to the entire human neocortex as the basis for man's primary and unique adaptation to his physical environment.

Neurognosis and Psychopathology

Let us now consider psychopathological entities that seem to involve not so much the gross structural relationships alluded to in chapter 3 and in the first part of this chapter but rather the microstructures or "holographic" models of reality discussed in chapters 4 and 5. Presumably it is these latter structures that are primarily responsible for alignment of "self,' either with the external world in general or with specific objects in particular.

The concept of "cognitive structure" is neither new nor revolutionary. In Beck's excellent discussions of the nature of depression (Beck, 1963, 1964), he describes a cognitive *structure* as a relatively enduring component of the cognitive organization in contrast to a cognitive *process,* which is transient. Other cognitive psychologists and psychiatrists have proposed concepts similar to "cognitive structure" to explain observed regularities of cognitive behavior. Thus Piaget's "schemata" (1948), Postman's "categories" (1951), Rapaport's "conceptual tools" (1951), Kelly's "personal constructs" (1955), the "coding systems" of Bruner et al. (1956), the "modules" of Sarbin et al. (1960), and the "concepts" of Harvey et al. (1961) are all conceptually very similar to "cognitive structure" as defined above. Many of these writers, however, use these terms either as "mental" structures or else as artificial constructs enabling one to conceptualize regularities in cognitive behavior. Even those who hint at a neural base for cognitive structures usually ascribe such neural configurations to the results of the organism's experience rather than to an interaction between experience and inborn pattern. Thus English and English (1958) have defined a cognitive schema (structure) as "the complex pattern, inferred as having been imprinted in the orgasmic structure by experience, that combines with the properties of the presented stimulus object or of the presented idea to determine how the object or idea is to be perceived and conceptualized." Provided one makes adequate provision for actual neural organization underlying such functions we are in total agreement with this definition. It corresponds with our concept of *acquired* models of reality. That such acquired models of reality do exist and, in fact, represent the individual organism's cognitive adaptive response to its particular experience is a position with which we would not argue. Where we do differ from most cognitive anthropologists and psychologists, however, is on the issue of imputing the *fundamental* neurognostic structures to an experiential base. The universal and adaptive qualities of the neurognostic structures described in chapter 5 (self-other dichotomy, causal thoughts, binary opposition, etc.) seem to mitigate against an experiential explanation for the formation of these neurognostic structures. Furthermore, Bower's evidence (1971) cited in chapter 3 strongly indicates that some structures a least are inherent. Granted that experiences may modify them, they are, nonetheless, too important to the organism's survival, and to the survival of the species, to credit their formation to the experience of each individual member of the species. We must emphatically exclude Piaget from this criticism, as he, more than others, has recognized the interactional component in the ontogeny of psychopathology.

We propose that much of the psychopathology that is manifested universally (i.e., insofar as it is independent of specific "cultural" determinants) can be explained either by inappropriate activation of a latent inherited neurognostic structure or else by a failure to deactivate such a structure when it is adaptive for it to be deactivated. The activation and deactivation mechanisms may be experiential or chemical, exogenous or endogenous. The essential point to be made here is that depressions, phobias, and obsessional characteristics may represent activation of specific inherited neurognostic structures at times when they are "normally" not activated. Let us first consider the issue of depression.

DEPRESSION

Beck's (1964) studies of depressed patients have forcefully pointed to the structural or schematic quality of depression. He notes:

> . . . as the depression deepened, there was a progressive dominance of the thought by these ideas (i.e., personal deficiency, self blame, negative expectations, etc.). Almost any external stimulus was capable of evoking a depressive idea (stimulus generalization) and depressive conclusions were drawn from the scantiest data. These processes contributed to a gradually increasing distortion and misinterpretation of reality.
>
> Some formal characteristics of the depressive cognitions were also noted. The conclusions, judgments, and interpretations appeared to rise automatically as though no prior reasoning was involved in their formulation. The cognitions had an involuntary aspect. Even when the patient was determined to suppress them or substitute other judgments he was unable to do so. (1964, p. 564)

Similar observed data with a great number of depressed patients reinforce the formulation that depression results from the activation of an inherited (but usually latent) neurognostic structure that aligns "self" to external reality in a negative fashion. All input is processed to some extent at least by the neurognostic structure so that even ordinarily positive circumstances are automatically seen in a negative light. This state continues until the structure is deactivated either chemically (by antidepressants) or perhaps by a consistent and prolonged modification of sensory input (e.g. psychotherapy).

In chapter 5 it was postulated that the evolution of a structure governing negative alignment of the self with external reality had the adaptive value of easing the transition of older dominant primates down the hierarchy. This would allow a relatively smooth transition of power without violent confrontations, which would threaten the foundation of primate group structure and thus endanger the species.

It is clear, however, that previously dominant animals moving down the hierarchy are not the only ones to appear withdrawn and depressed. McKinney, et al (1972) and others have shown that primates display depressive-like behavior in a variety of circumstances. Perhaps most notable is the "depression" seen in young animals separated from their mother. Similar reactions in humans have been described by Spitz (1945), Spitz and Wolf (1946), Micic (1962), Schaffer and Callender (1959), and Robertson (1970). Their work tends to confirm that the structures governing alignment of self to the environment (whether positive or negative) apparently evolved in relation to the issue of mastery over the environment. An infant "masters" its environment via its symbiotic relationship with its mother. If the mother is removed, a few failures (inability to "master" the environment) activates the neurognostic structure that governs a negative alignment to the environment. The animal then becomes "depressed," withdrawn, and quickly dies, making little or no effort to use whatever abilities it does have. Put in a rather brutal "survival of the fittest" context, any animal whether infant, aging dominant male, or whatever, that cannot exercise some minimal degree of mastery over his environment is a hindrance to the primate group. Since an efficient group structure is essential to the survival of individual group members, the neurognostic structure governing "depression" and withdrawal has strong evolutionary significance. It simply eliminates "parasitic" members of the group. This may be through the withdrawal and death of a motherless infant, or through the relinquishing of a dominant position by an animal no longer capable of fulfilling it. Seen in this light, the normally latent neurognostic structure controlling depression has as much evolutionary significance as the structure governing a positive alignment of self to the environment. Although the latter gives a feeling of mastery and results in assertive positive action, the former maintains optimal role functioning within a group structure by removing group members from roles they can no longer efficiently fill.

Such an evolutionary perspective helps us understand the depression engendered in human infants by separation from their mothers during certain critical periods. Bowlby (1969) and later Rivinus and Katz (1971) have presented evidence that mother-infant bonding represents a complex interdigitated triggering of response on both parts. This complex interaction is probably biogenetically structured and represents, for the infant, the first glimmerings of mastery over the environment. Separation undoubtedly triggers the reverse response, also biogenetically structured, implying loss of mastery with concomitant depression. The issue of loss of mastery over the environment can also account for the marked prevalence of depression among the aged.

There are certainly many other circumstances that can and do activate the depressive structure in man. Unless the activating cause is primarily chemical, however, we contend that analysis of the precipitating circumstances will almost always reveal the ancient adaptive issues of lack of control or mastery in some context or other. Even when the precipitating cause *is* a chemical abnormality, we maintain that this abnormality activates the depressive structure. For example, Redmond (1973) has analyzed the behavior of free-ranging macaque monkeys after injection of 6-hydroxydopamine into their cerebral ventricles. This substance causes a depletion in catecholamines, a lack of which has been widely implicated in human depression. Observers noted that the treated animals fell in the dominance hierarchy; previously dominant animals became subordinate and even peripheral to the group.

ALCOHOLISM

Biogenetic structuralism can also give us an insight into one of the probable causes of problem drinking. If indeed human beings carry within the brain models of reality that are either confirmed by sensory input or modified by it, one must maintain that this is a biologically important and highly adaptive process. Assuming that this is true, one must postulate a physiological imperative to match sensory input with neurognostic models. This is an aspect of the cognitive imperative that was alluded to earlier in this book and elsewhere (d'Aquili, 1972). A corollary of this position is that the frustration of this imperative, i.e., the chronic inability to modify appropriately neurognostic models according to sensory input, must be dysphoric. One can speculate that the clouding of the sensorium produced by the ingestion of alcohol decreases the perception of the discrepancy between the expectation of environmental events (based on unmodified neurognostic models) and unexpected sensory input. Thus a clouding of the awareness of this discrepancy would be expected to produce a relief from the chronic dysphoria (anxiety/depression) generated by an inability to satisfy adequately the cognitive imperative. As this book was going to press, the authors became aware of some evidence supporting this position. Rodgers (1972) has independently proposed a similar model for alcoholism based on the Miller, Galanter, and Pribram (1960) TOTE mechanism. The TOTE theory of behavior was alluded to during our discussion of neurognostic models and represents a model modification system similar to that espoused in this book. Using the TOTE system, Rodgers (1972) states:

> The assessment of absence of significant discrepancy is obviously made by neurological comparisons within the central nervous system. I suggest that alcohol pharmacologically reduces the precision of this comparison, such that a goal can be perceived as having been achieved while there is a greater discrepancy between cognitive models and sensory evidence than is permissible in a "normal" state. As a minor example, a person with an alcohol load can perceive his foot as having reached a satisfactory weight-bearing position during walking before it in fact is ready to receive a transfer of weight. This simple misperception can result in the stagger of the intoxicated subject. More significantly, all motivated behavior is directed toward achieving desired goals, i.e., toward eliminating perceived discrepancy between a cognitive model and the "real world." Such perceived discrepancy is normally eliminated by hard work directed towards changing the real world. To the extent that alcohol reduces the perceived discrepancy pharmacologically, it subjectively substitutes for hard work. The substitution is illusory, of course, since the real world is not brought into conformity with subjective desire but is merely perceived as having been brought into conformity, and only for the duration of the pharmacologic state. (p. 223)

It seems obvious that such a model is identical with the one generated by biogenetic structuralism with a few minor differences concerning the nature of neurognostic models. We feel that this general approach, and research devised to test it, will yield a rich harvest of information concerning the basic motivation of much problem drinking. We would also suggest that chronic inability to modify adequately neurognostic models by novel sensory input clearly puts the subject at an adaptive disadvantage. Theoretically this eventually ought to activate the structure governing a negative alignment between him and the environment. In more familiar language the perception (chronically) of mismatch between expectations and what actually occurs in the real world ought to trigger an anxiety-depression syndrome. Seen in this light the ingestion of alcohol can be interpreted as a defense against overwhelming depression owing to a lack of "cognitive mastery" of the environment. Of course, chronic alcoholism can produce such social and familial disruption in a person's life that severe depression can be precipated because of an actual loss of mastery over real events. Thus a vicious cycle may be formed. An individual drinks to avoid the dysphoric state generated by a lack of "cognitive mastery." The "cure" however eventually so disrupts his life as to produce a major loss of control over external events (family, job, etc.). This in turn precipates an even more severe depression. It has been known for some time that alcoholism and depression are very closely related, probably on many levels. Studies aimed at elucidating the genetics of both alcoholism and depression (Pitts and Winokur, 1966; Winokur and Clayton, 1968; Winokur et al., 1970, 1971; Schuckit et al., 1969; Schuckit, 1972; Goodwin, 1971, 1973) have strongly indicated that this relationship is at a more basic level than simply depression secondary to the social disruption of chronic drinking. We feel that the biogenetic structural interpretation of alcoholism and depression provides at least one more level upon which the two entities may be related and points the way for future empirical investigation.

PHOBIA

The biogenetic structural approach may help us to understand most if not all severe clinical phobias. In chapter 5 we discussed the possibility of inborn structures governing either a positive or negative alignment of the "self" toward certain *specific* objects in the external environment. Such alignments, being governed by inherited neurognostic structures, would obviously have adaptive significance to the individual. The negative alignment towards objects that may have represented danger to survival in man's ancient past we have equated with the clinical entity called phobia. Seligman (1971) has convincingly argued that classical phobias, which are usually presented to a psychiatrist for treatment, differ from the usual conditioned fear response both in terms of the number of pairings of stimulus and fear response necessary to establish the relationship and in terms of the difficulty of extinguishing the response. In a critique of what has become the accepted learning theory explanation of the genesis of phobias, Seligman writes (p. 312):

> According to Pavlov's view of conditioning the choice of CS (conditioned stimulus) is a matter of indifference. "Any natural phenomenon chosen at will may be converted into a conditioned stimulus . . . any visual stimulus, any desired sound, any odor and the stimulation of any part of the skin" (Pavlov, 1928, p. 86). This is the heart of the general process view of learning and, by this widely held view, any CS which happens to be associated with trauma should become phobic. But a neglected fact about phobias is that, by and large, they comprise a relatively nonarbitrary and limited set of objects: agoraphobia, fear of specific animals, insect phobias, fear of heights, and fear of the dark, etc. All these are relatively common phobias. And only rarely, if ever, do we have pajama phobias, grass phobias, electric-outlet phobias, hammer phobias, even though these things are likely to be associated with trauma in our world. The set of potentially phobic events may be nonarbitrary: events related to the survival of the human species through the long course of evolution.

After reviewing recent literature relative to the "preparedness" (chapter 5) of phobias, he concludes (p. 312):

> Phobias are highly prepared to be learned by humans, and, like other highly prepared relationships, they are selective and resistant to extinction, learned even with degraded input, and probably are noncognitive.

From this it seems reasonable to conclude that the most profitable way of looking at the genesis of phobias is from a neurostructural perspective. A phobia develops when a previously latent neural structure governing a negative alignment of the "self" towards an object becomes activated. Such an object is best viewed as having posed a threat to survival during the long period of hominid evolution. The selective advantage in the development of such neurognostic structures is obvious. It is tempting to speculate that perhaps the reason such phobias are not present in every member of our species is a reflection of the fact that the development of culture no longer made the threat from these objects immediate. In other words, it may no longer represent as strong a selective advantage to maintain these neurognostic structures. They may thus be slowly dropping out of the population. Particularly if they are polygenically determined, however, one would expect this process to be slow indeed.

OBSESSIVE-COMPULSIVE TRAITS

More speculatively than is the case with either depression or phobias, one can consider certain aspects of obsessive-compulsive behavior within a neurostructural framework. In chapter 5 we considered briefly the cognitive imperative, i.e., the drive in man to organize events presented in the external world within a cognitive matrix. We have contended that the cognitive matrix is made up of a number of inherited neurognostic structures, most prominent of which are those that govern causal relationships, and those that govern "binary opposition" (whether these latter relationships are antinomous or comparative in nature). Since one of the major aspects of the obsessive-compulsive personality is the intense desire to arrange thoughts and environmental circumstances in an exceedingly orderly and precise fashion, may not one ask whether a defect in activation or deactivation of the neurognostic structures described above is involved in this psychopathology? It may be that in the obsessive-compulsive personality there are a number of stimuli regularly activating these structures that do not do so in the "normal" person. In other words, these neurognostic structures may be activated by a much wider range of stimuli in the obsessive-compulsive personality than in the "normal" individual. An alternative mechanism (and perhaps a more probable one) would be that all sensory input is processed by these structures to some extent at least. In the "normal" person there may be selective deactivation of the structures depending on the importance of the information to the organism. The obsessive-compulsive may not be able selectively to deactivate these neurognostic structures, and all input, of whatever importance, is rigidly and continuously processed by them.

In any case, we present this formulation of obsessive-compulsive behavior not as fact but as the basis for possible future investigation.

CRITICAL PERIODS

Let us now devote a few words to a more general explication of the concept of "critical periods." As we have seen (chapter 7), the term may be used quite explicitly to define characteristics of linguistic or communicative ontogeny in man and other animals. But the term actually has a broad usage in biology. In embryology, for example, the concept refers to time periods of maximum sensitivity or indifference to chemical forces acting on the cellular mass. Connolly (1972) has reviewed the subject in relation to learning and developmental psychology in general. In this context a critical period can be viewed as a period of maximum sensitivity to certain kinds of learning. If the organism is not subjected to appropriate input during a given critical period, it may suffer a more or less permanent deficit in the area of learning which that particular critical period influences. Thus "imprinting" in Lorenz's (1935) work with birds represents learning during a critical period. Sluckin (1964) and Bateson (1966) have both extensively reviewed the literature on imprinting. Unquestionably there is good evidence for the existence of sensitive periods for the establishment of primary social bonds in many species. Bowlby (1969), Yarrow (1960), and Klaus et al. (1972) provide good evidence that the same situation may be true in humans.

Needless to say, the concept of critical periods is not confined to periods governing the formation of primary social bonds. It is increasingly felt that much of the advancement of cognitive and affective learning in the development of an individual takes place during critical or sensitive periods that in turn reflect progressive maturation or readiness of the central nervous system. Systems such as Piaget's are, of course, based on such a model, as is that of Lenneberg for language acquisition (see chapter 7).

We bring up this topic at this point because we feel that biogenetic structuralism can offer a new perspective on the concept of critical periods. It has been traditionally maintained that a critical period involves the optimal time for certain kinds of learning (understood in a classical sense). Insofar as such learning involves setting up acquired models through neurognostic structure, the concept of critical period is perfectly compatible with the theory of biogenetic structuralism presented in this book. Indeed one would expect from the theory some sequential outlay of critical periods for learning on the basis of progressive maturation and development of the central nervous system. In biogenetic structuralist terms, however, the concept of critical period gains an added perspective. Thus a critical period of psychological and/or cognitive development can be viewed not only as a period for the formation of acquired models of reality (via classical learning mechanisms) but also as the optimal period for activating and developing (elaborating) specific inherited neurognosis and neurognostic structures.

As we have said, our purpose is not to present a comprehensive scheme of psychological and/or cognitive development. We again refer the reader to the works of Piaget, and Harvey, Hunt, Schroder, and their associates. Nor is it our intent to explain psychopathology solely in terms of critical periods. What we do hope to emphasize, however, is that a neurostructural approach offers a valuable (indeed, necessary) perspective in *these* areas as well as in the areas of anthropology and linguistics. Aspects of problems, heretofore very puzzling, become more amenable to analysis when one applies an evolutionary neurostructural approach. We can only hope that future investigators will more comprehensively study neurostructural aspects of psychopathology, psychology, and cognitive development.

References

d'Aquili, E., et al. 1966. "Reflex Resistance to Anaesthesia in Partially Denervated Neurons." *Experimental Neurology*, 16:307-315.

——1972. *The Biopsychological Determinants of Culture.* Addison-Wesley Modular Publications.

Bateson, P.P.G. 1966. "The Characteristics and Context of Imprinting." *Biological Reviews,* 41:177.

Beck, A.T. 1963. "Thinking and Depression: 1. Idiosyncratic Content and Cognitive Distortions." *Archives of General Psychiatry,* 9:324-333.

——1964. "Thinking and Depression." *Archives of General Psychiatry,* 10:561-571.

Bleuler, E. 1950. *Dementia Praecox, or the Group of Schizophrenia.* New York: International Universities Press.

Bonin, G., and P. Bailey. 1947. *The Neocortex of Macaca Mulatta.* Urbana: the University of Illinois Press.

Book, J.A. 1953. "A Genetic and Neuropsychiatric Investigation of a North Swedish Population." *Acta Genetica et Statistica Medica,* 4(1).

——1960. "Genetical Etiology in Mental Illness." *Milbank Memorial Fund Quarterly,* 28, July 3.

Bower, T.G.R. 1971. "The Object in the World of the Infant." *Scientific American,* October (225)30.

Bowlby, J. 1969. *Attachment and Loss,* Vol. 1. New York: Basic Books.

Bruner, J.S., J.J. Goodnow, and G.A. Auston. 1956. *A Study of Thinking.* New York: John Wiley.

Connolly, Kevin. 1972. "Learning and the Concept of Critical Periods in Infancy." *Developmental Medicine and Child Neurology.* 14:705-714.

Cowie, V.A., A. Coppen, and P. Norman. 1960. "Nuclear Sex and Body Build in Schizophrenia." *British Medical Journal,* 2:431-433.

English, H.B., and A.C. English. 1958. *A Comprehensive Dictionary of Psychological and Psychoanalytical Terms.* New York: Longmans, Green and Company.

Fish, B. 1961. "Study of Motor Development in Infancy, and Its Relation to Psychiatric Functioning." *American Journal of Psychiatry,* 117:113-118.

Goldfarb, W. 1964. "An Investigation of Childhood Schizophrenia." *Archives General Psychiatry,* 11:620-629.

Goodwin, Donald W. 1971. "Is Alcoholism Hereditary?" *Archives of General Psychiatry,* 25:545-549.

Goodwin, Donald W., et al. 1973. "Alcohol Problems in Adoptees Raised Apart from Alcoholic Biological Parents." *Archives of General Psychiatry,* 28:238-243.

Gottesman, I., and J. Shields. 1967. "A Polygenic Theory of Schizophrenia." *Proceedings of the Academy of Sciences,* U.S.A. 58:199-205.

Harvey, O.J., D.E. Hunt, and H.M. Schroder. 1961. *Conceptual Systems and Personality Organization.* New York: John Wiley.

Hertzig, M., and H. Birch. 1966. "Neurologic Organization in Psychiatrically Disturbed Girls." *Archives of General Psychiatry,* 15:6.

——1968 "Neurologic Organization in Psychiatrically Disturbed Adolescents." *Archives of General Psychiatry,* 19:528-537.

Heston, L.L. 1966. *British Journal of Psychiatry.* 112:819.

——1970. "The Genetics of Schizophrenic and Schizoid Disease." *Science,* 167:248-255.

Heston, L.L., and D. Denny. 1968. "Interactions Between Early Life Experience and Biological Factors in Schizophrenia." In D. Rosenthal and S. Kety (eds.), *The Transmission of Schizophrenia,* New York: Pergamon Press.

Hinde, R.A. 1969. *Bird Vocalizations.* Cambridge: Cambridge University Press.

Holloway, R.L., Jr. 1966. "Cranial Capacity, Neural Reorganization, and Hominid Evolution: A Search for More Suitable Parameters." *American Anthropologist,* 68:103-121.

Huxley, J., et al. 1964. "Schizophrenia as a Genetic Morphism." *Nature* (London), 204:220, October 17.

Kallman, F.J. 1938. *The Genetics of Schizophrenia.* New York: Augustin.

——1946. "The Genetic Theory of Schizophrenia: An Analysis of 691 Schizophrenic Twin Index Families." *American Journal of Psychiatry.* 103:309-322.

Kaplan, A.R., 1965. "On the Genetics of Schizophrenia." *Eugenics Quarterly,* 12, 132-136, September.

Kelly, G.A. 1955. *The Psychology of Personal Constructs,* Vol. 1. New York: W. W. Norton.

Kety, S., D. Rosenthal, F. Wender, and F. Schulsinger. 1968. "The Types and Prevalence of Mental Illness in the Biological and Adoptive Families of Adoptive Schizophrenics." In D. Rosenthal and S. Kety (eds.), *The Transmission of Schizophrenia.* New York: Pergamon Press.

Klaus, M.H., R. Jerauld, N. C. Kreger, W. McAlpine, M. Steffa, and J. H. Kennell. 1972. "Maternal Attachment. Importance of the First Post-Partum Days." *New England Journal of Medicine,* 286, 460.

Lancet. 1966. *Genetics of Schizophrenia,* 2:692, September.

Lorenz, K. 1935. "Der Kumpen in der Umwelt des Vogels." *Journal of Ornithology,* 83, 137, 289.

Luria, A.R. 1966. *Higher Cortical Functions in Man.* New York: Basic Books.

Marler, P. 1970. "A Comparative Approach to Vocal Learning." *Journal of Comparative and Physiological Psychology,* 71(2, part 2):1-25.

McKinney, W.R., et al. 1972. "How They're Using Monkeys to Study Depression." *Resident and Staff Physician,* February.

Micic, Z. 1962. "Psychological Stress in Children in Hospital." *International Nursing Review,* 9:23-31.

Moran, P.A. 1965. "Schizophrenia as a Genetic Morphism." *Nature,* (London), 206:1113, June 12.

Miller, G.A., E. Galanter, and K.H. Pribram. 1960. *Plans and the Structure of Behavior.* New York: Holt.

Nottebohm, F. 1972. "Neural Lateralization of Vocal Control in a Passerine Bird: II. Subsong, Calls, and a Theory of Vocal Learning." Journal of Experimental Zoology, 179:35-50.

Pavlov, I.P. 1928. *Lectures on Conditioned Reflexes.* New York: International Publishers.

Piaget, J. 1948. *The Moral Judgment of the Child.* Translated by M. Gabain. Glencoe, Ill.: The Free Press.

Pitts, F.N., Jr., and G. Winokur. 1966. Affective Disorders. VII. Alcoholism and Affective Disorder." *Journal of Psychiatric Research.* 4:37-50.

Postman, L. 1951. "Toward a General Theory of Cognition." In J.H. Rohrer and M. Sherif (eds.), *Social Psychology at the Crossroads.* New York: Harper.

Prout, C.T., and M.A. White. 1956. *Journal of Nervous and Mental Disease.* 123, 162.

Rapaport, D. 1951. *Organization and Pathology of Thought.* New York: Columbia University Press.

Redmond, D.E. 1973. "Behavior of Free-Ranging Macaques After Intraventricular 6-Hydroxydopamine." *Science,* 181:1256-1258.

Rivinus, H.A., and S.H. Katz. 1971. "Evolution, Newborn Behavior, and Maternal Attachment." *Comments on Contemporary Psychiatry,* 1, 3:95-104.

Robertson, James. 1970. *Young Children in Hospital* (2nd edition with a postscript). London: Tavistock Publications. New York: Barnes and Noble.

Rochford, J., et al. 1970 "Neuropsychological Impairments in Functional Psychiatric Diseases." *Archives of General Psychiatry,* 22, February.

Rodgers, D.A. 1972. "A Psychological Interpretation of Alcoholism." In *Nature and Nurture in Alcoholism, Annals of the New York Academy of Sciences* (ed. by F.A. Seixas, G.S. Omenn, and E.D. Burk), Vol. 197.

Rosenthal, D. 1968. "The Heredity-Environment Issue in Schizophrenia: Summary of the Conference and Present State of Our Knowledge." In D. Rosenthal and S. Kety (eds.), *The Transmission of Schizophrenia.* New York: Pergamon Press.

Sarbin, T.R., R. Taft, and D.E. Bailey. 1960 *Clinical Inference and Cognitive Theory.* New York: Holt, Rinehart and Winston.

Schaffer, H.R., and W.M. Callender. 1959. "Psychological Effects of Hospitalization in Infancy." *Pediatrics,* 24:528-539.

Schuckit, Marc A. 1972. "Family History and Half-Sibling Research in Alcoholism." In *Nature and Nurture in Alcoholism, Annals of the New York Academy of Sciences* (ed. by F.A. Seixas, G.S. Omenn, and E.D. Burk), Vol. 197.

Schuckit, M., F.N. Pitts, Jr., T. Reich, L.J. King, and G. Winokur. 1969. Alcoholism. I. Two Types of Alcoholism in Women. *Archives of General Psychiatry,* 20:301-306.

Seligman, M. 1971. "Phobias and Preparedness." In *Behavior Therapy,* 2, 307-320.

Slater, E. 1958. "The Monogenic Theory of Schizophrenia." *Acta Genetica et Statistica Medica,* 8(1):50-56.

Sluckin, W. 1964. *Imprinting and Early Learning.* London: Methuen.

Spitz, R.A. 1945. "Hospitalism." *This Anual,* 1:53-74.

Spitz, R.A., and K.M. Wolf. 1946. "Anaclitic Depression." *This Annual,* 2:313-342.

Winokur, G., R. Cadoret, J. Dorzab, and M. Baker. 1971. "Depressive Illness: a Genetic Study." *Archives of General Psychiatry,* 24:135-144.

Winokur, G., and P. Clayton. 1968. "Family History Studies. IV. Comparison of Male and Female Alcoholics. *Quarterly Journal of Studies on Alcohol.* 29:885-891.

Winokur, G., T. Reich, J. Rimmer, and F.N. Pitts, Jr. 1970. "Alcoholism. III. Diagnosis and Familial Psychiatric Illness in 259 Alcoholic Probands." *Archives of General Psychiatry,* 23:104-111.

Yarrow, L.J. 1960. "Maternal Deprivation: Toward an Emprical and Conceptual Re-evolution." *Psychological Bulletin,* 58:459.

IX

Implications for Social Science

For a biologist, the term "culture" comes to mean a set of behavior characterized by its *origin*. An individual develops a particular behavior partly because its genetic endowment directs its development, and partly because the environment feeds information into the process of development. If a behavior were entirely programmed by genes, it could be called "innate"; if it were an exclusive product of environmental stimuli during the animal's ontogeny, it could loosely be called "acquired" behavior. In reality, these extremes do not occur.—HANS KUMMER, *Primate Societies*

LET US SUMMARIZE the tenets of biogenetic structuralism and then proceed to a final discussion of the ramifications of the position for American social science. In the preceding chapters we have developed arguments for the following claims:

(1) That there is no level of reality intervening between *Homo sapiens* as a biological phenomenon and that organism's environment. In other words, human behavior is the result of a dialectic between the central nervous system, primarily the higher cortical functions, and the environment. All other asserted or posited levels of reality have analytic status only.

(2) That the behavior we are used to calling "cultural" and "purely learned" is actually variant behavioral manifestations of universal structural models. These models are probably phenotypic

expressions of dendritic-synaptic connections located within the brain. Structural configurations are initially totally, and later partially, determined by genetic coding and are susceptible to the pressures of natural selection.

(3) The modes of "thought," "reason," "cognition," "sciencing," "mythologizing," "magical causation," and the like are actually the behavioral equivalents of internal, neurophysiologically structured, and systematic channels of sensory association and processing characteristic of the human brain, as well as of the brains of other organisms.

(4) That most of the biogenetically determined cognitive processing occurring in the human brain takes place outside the bounds of human awareness—the latter phenomenon being more an experiential by-product of cognitive processing than an entity isomorphically independent of cognitive processing.

These conclusions are controversial and will no doubt be attacked from some quarters as "metaphysical," "reductionistic," or even "fantasy." Behaviorists who define "learning" solely in terms of input-output (see chapter 4), and who have until quite recently treated the brain as a black box, may argue that no direct evidence exists of any such entities as holographic or neurognostic models or empirical modification cycle. American anthropologists, as yet relatively unmoved by the principal formulations and insights of the French structuralist school (notably those of Piaget and Levi-Strauss), may react by reasserting the reality of the absolute distinction between human institution and behavior (which are "learned") and animal behavior (which is inherited)—in other words, by dredging up the long-standing, although often dormant, nature-nurture dispute.

As we have shown, learning cannot be understood apart from either preparedness *or* input. This conception of learning has a number of important consequences for social science in general and for anthropology in particular. We will consider here these consequences in relation to the concept of "culture" and will therefore address our final remarks precisely to this notion as traditionally understood. Much of what follows is a critique of a major trend in cultural anthropology, since this is the science of "culture" par excellence. But, to a greater or lesser extent, it is a critique of the social sciences in general since the objects of their study are usually reified entities either derived from the traditional concept of "culture" (e.g., "enculturation," "acculturation"), or else included within the concept (e.g., "class," "values," "institution"). Biogenetic structuralism, therefore presents a challenge not only to anthropology but also to all the social sciences insofar as "culture" and related reified concepts are called into question.

Culture

The concept "culture" has been variously defined. Two American anthropologists, Kroeber and Kluckhohn (1963), devoted an entire monograph to examining and evaluating over a hundred definitions of culture, and they found them all wanting. They did indicate, however, two common themes running through most formulations of the term: that culture is *learned* and that it is *shared* by members of any society. These two principal attributes of the culture concept remain with us to this day, as any cursory examination of modern introductory anthropology texts will indicate: "Members of human societies always share a number of distinctive modes or ways of behaving that, taken as a whole, constitute their culture. Each human society has its own culture, distinct in its entirety from that of any other society" (Beals and Hoijer, 1971, p. 103). "Cultures are patterns of behavior, thought, and feeling that are acquired or influenced through learning and that are characteristic of groups of people rather than of individuals" (Harris, 1971, p. 136).

It is the nineteenth-century anthropologist, E. B. Tylor, who is commonly given credit for having first defined the concept of culture, and many anthropologists today feel that his formulation remains viable. The foregoing two themes, *learning* and *sharing*, are certainly inherent in his notion of culture, but what is usually missed in relating the history of the concept is that Tylor was the last social theorist having strong influence upon American ethnological thinking who was capable of entertaining a view of man as simultaneously cultural, biological, and structural.,

It was Tylor who could picture human society as being comprised of almost outlandish elaborations in art, music, religion, ritual, etc., as well as patterns of behavior and attitude essentially functionless in relation to the total social system (i.e., his notion of "survivals"). Yet for Tylor man was every bit a biological phenomenon. The major function of human society he saw as economic, the production of the goods and services required for biological survival. The general style of cultural elaboration manifest in any particular society was ultimately determined by the stage of evolutionary development characteristic of that society. Finally, Tylor saw that man exhibits certain structural regularities on a global level, which he offered as evidence of the "psychic-unity" of man, the proclivity of men at any given stage of development to come to similar solutions to similar problems.

Because of American reaction to both the evolutionist and the "psychologistic" themes in nineteenth century social philosophy, American anthropologists embraced the culture concept espoused by Tylor to the virtual exclusion of the other formulations central to his thinking. Thus Tylor left a truncated legacy. And American anthropology was led astray—into an intense concern for precisely those aspects of human behavior least likely to produce a truly nomothetic science of man. Thus Tylor's legacy was even further distorted by scholars from such diverse theoretical positions as Kroeber (1948) and White (1959). The former explicity stated that culture is a "superorganic" phenomenon dependent on biology only insofar as the "culture-bearers" are dependent on biological principles for survival. Otherwise culture has a life and internal processes of its own. Leslie White (1959) returned to nineteenth century cultural evolutionary theory in many respects, but with the difference that there is hardly a nod in the direction of biology. Culture develops and unfolds according to its own laws and represents an internally consistent and closed system insofar as biological variables are all but ignored. At this point, it is clear that culture has become "culture," a reified global entity. So extreme are these positions that they have been called "culturology."

We do not wish to imply that White's or Kroeber's positions are accepted by all American anthropologists. Far from it! But so widely diffused has the notion of superorganic culture become (among both laymen and professionals alike) that one hears it spontaneously even from the lips of scholars who repudiate the notion when explicitly questioned on that point. In other words, "culturology" has had an insidious and pervasive influence on anthropology in particular, and to a great extent on the social sciences in general.

The Demise of "Culturology"

There are at least three reasons that the concept of culture must be radically revised if it is to retain any explanatory significance at all in the social sciences. In the first place "culture," understood in a reified sense, has become obsolete and will likely lapse into disuse. In this sense, the concept is commonly given epiphenomenal expression. What begins as a deceptively handy shorthand for central tendencies in behavior, interaction, or attitude in a society quickly becomes converted conceptually to another "level of reality." Once the analytically derived "culture" is reified and given the ontological status of perceived reality, it constitutes a conceptual barrier to explanations of social behavior grounded in biological processes. Biological explanations of "cultural" phenomena are termed reductionistic and are treated as fallacious.[1] Therefore, in "culture" we are faced with the fallacy of misplaced concreteness. The result of such reasoning renders cultural theory incapable of accounting for the important and dialectic relationship between brain and environment. This relationship becomes crucial in biogenetic structuralist theory, where a meaningful concept of culture encompasses such constructs as "latent processes," "latent structures," "neurognostic models," all of which, as we have seen, refer to genetically structured association processes within the brain, and their modification, modulation, and elaboration through interaction with environment.

[1] French anthropology and British anthropology are as much at fault here as American anthropology.

The second reason for the demise of "culturology" has been already suggested. "Culture" as a reified concept tends to bias our reasoning in favor of a fundamental bifurcation between *Homo sapiens* and all other species of organisms. As Hallowell (1960) suggests:

> . . . this preoccupation with culture led to a *re*creation of the old gap between man and other primates which, it was thought, the adoption of an evolutionary frame of reference would serve to bridge. The repeated emphasis given to speech and culture as *unique* characteristics of man sidestepped the essence of the evolutionary problem. Distinctive characteristics of the most highly evolved primate were asserted without reference to prior capabilities, conditions, and events in the evolutionary process that made this characteristic mode of adjustment possible. For unless culture and speech be conceived as sudden and radical emergents, they must be rooted in behavioral processes which can no more be considered apart from the general framework of behavioral evolution than the distinctive structural characteristics of man can be considered apart from morphological evolution. (p. 319)

"Culture," in a word, is an anthropocentrism of the first order. Considering some of the prevailing definitions of culture, the built-in assumption is that the preponderance of man's behavior is learned and that of all other species is at least primarily instinctual. Nowhere is this more evident than in the study of the origin and acquisition of language. Until quite recently, anthropological linguists have been forced to define "language" by increasingly fine distinctions in order to retain the dichotomy between human speech and the communication systems of other organisms. Hockett (1960) in a famous article isolated thirteen design features of communication systems. Of these, only one, duality of patterning, he found to be uniquely characteristic of human speech. More recently, Hockett (Altmann and Hockett, 1968) and others (Gardner and Gardner, 1969; Premack, 1971; Hill, 1972) have been forced to the conclusion that natural communication systems differ more by degree and superficial detail than by essential or structural quality —a fact we have seen emerging in the previous chapters.

Prior to the mid-sixties, and with authors acting under the presuppositions generated by the Whorf-Sapir brand of cultural/linguistic relativism, language acquisition studies were concerned only with the geographical and cross-linguistic distribution of surface aspects of language (Braine, 1963; Brown and Fraser, 1963), or in grammatical feature studies (Klima and Bellugi, 1966; Brown and Bellugi, 1964). It was only after the publication of Slobin's work (1966, 1968) on language acquisition among Russian children that researchers began to pay attention to universal structures of language.[2] It has since become evident in the research of language origin and acquisition specialists (Lenneberg, 1967; McNeill, 1971; Slobin, 1971) that the basic structures upon which language is built by the child are biogenetically given and universal (see chapter 7).

To continue to perpetrate the conceptual schism between man, the so-called culture-bearer, and the rest of the animal kingdom as predominantly instinct-driven is to burden scientific reason with imponderables. As the poet, Tom Sullivan, once said, "in man the universe becomes aware of itself." Is it not ironic that in becoming aware, man even as social scientist, proceeds to abstract himself conceptually from the universe? Ironic, yes—but also understandable—, considering that social scientists are men and that men are able to introspect and report the results of their internal experience. It is this capacity for subjectivity, juxtaposed with the fact that we have no subjective reports from other organisms, that bends our reason to anthropocentrism. At any rate, biogenetic structuralists assume that essential similarities exist between *Homo sapiens* and other members of the animal kingdom, especially with other mammals. In order to account for the origin and behavior of *Homo sapiens*, similarities are often more critical than differences.

An attack on this question has been made from another quarter —from students of primate ethology. A number of scholars have indicated or implied that by any reasonable, nonanthropocentric definition of culture, certain nonhuman primates must be viewed as culture-bearers (Mann, 1972; Sade, 1968; Jolly, 1972; Kummer, 1971; Chance and Jolly, 1970). "A great deal of the behavior of the primates can be called cultural, in the sense that it is transmitted by learning from generation to generation. This is true not only of social behavior but of behavior toward the environment, as simple a thing as the traditional home range of a troop" (Jolly, 1972; p. 350).

[2] For a concise history of language acquisition studies, see Isbell (1972).

Mann (1972) has demonstrated that the definition of culture offered by Goodenough applies equally well to learned behavior in primate societies:

> Goodenough . . . has suggested a definition of culture which includes those things that are learned and those things that one needs to know in order to meet the standards of others. Culture can be seen as standards of appropriate behavior, learned by the individual, which allow him successfully to interact within his own social and group milieu. It is apparent that this definition also applies to the social system of the non-human higher primates. During the socialization period most group-living primates learn acceptable behavior, practiced in peer play-groups, which permits them to interact successfully within the troop. (p. 280)

The third and most important reason for the demise of a reified concept of "culture" is that this concept, and all it logically entails, tends to direct our attention to the exotic, the superficial ways by which the members of one society are distinguished from members of another society. Remembering again the two commonly held attributes of the culture concept, we note the stress upon *shared* patterns, the emphasis, of course, being upon intrasocietal, not intersocietal, sharing. It was Henry A. Murray who remarked that in some ways no men are alike, in some ways some men are alike, and in some ways all men are alike. Psychology has variously centered on the ways no men are alike (differential psychology), some men are alike (personology), and in an ethnocentric sense, the ways all men are alike (learning psychology). Cultural anthropology has primarily concerned itself with the ways some men are alike. We do not deny the importance and validity of such concerns. They have contributed immeasurably to our understanding and appreciation of the rich diversity that man exhibits. But unless a study of human behavioral (cultural) differences is based upon a unifying and valid theoretical framework, it is no more than a sort of natural history with no claim to full scientific status. Cultural anthropology may perhaps call itself a science insofar as it is a systematic ordering of facts. But neither it, nor most of the other social sciences, can be considered sciences in the sense of being organized bodies of knowledge, based on general principles, and in constant "empirical dialogue" with the theoretical foundations upon which they rest. The anthropological study of the ways all men are alike has, within this century, been only within the purview of physical anthropology and modern French structuralism.

Because of its concern for the exotic, cultural theory has in some ways inhibited the emergence of a nomothetic social science. To put this in pragmatic terms, anthropology, for example, remains a pre-paradigmatic science. It is characteristic of what Kuhn (1970) has called prenormal science that its practitioners have no single universal conceptual scheme by means of which to evaluate their individual results. The discipline is fractionated and factionalized. Practitioners are, in the main, involved in the particular. And so it is with most of the social sciences, which today are in a state analogous to that of the field of chemistry prior to the advent of the periodic table. Prior to the formulation of that now justly famous model, no single theory existed by means of which researchers working on the behavior of sodium could meaningfully relate their results to those of researchers working on potassium. It seems indicative to us that anthropology continues to categorize cultural anthropologists by geographical ("culture") area, e.g., "Africanists," "Latin Americanists."

The concept of a reified "culture" has been mistakenly characterized as an anthropological paradigm. This is by no means the case. A paradigm, to paraphrase Kuhn, consists of a single, all pervasive theory or coherent body of theory that accounts, at least in part, for a majority of the phenomena of concern to the practitioners of a particular discipline. Adherence to the major tenets of the paradigm is requisite for membership in the discipline. Furthermore, the paradigm defines for the members problems relevant for research and the proper methodology for carrying out that research. For a member of such a discipline to examine a problem defined in the paradigm as irrelevant or contradictory to the paradigm is to commit a breach of scientific discipline that borders on heresy. No young geneticist on his way up, for example, would choose for his doctoral dissertation topic a replication of Paul Kammerer's experiments on the inheritance of acquired characteristics.

This view of "normal" science clearly does not describe the

state of most of the social sciences. Concepts like "culture," "society," "structure," and "role" no more constitute paradigms in the social sciences than the concepts "mass," "electron," "dendrite" or "reagent" constitute paradigms in other scientific disciplines. Nor does the so-called process school of archaeology offer that subfield a new paradigm. All the "new" archaeology amounts to is a rediscovery by archaeologists of induction. Many of us have said to our students "you can do anything and call it social science." We are able to say this with some accuracy simply because there are no paradigms to which we all owe our allegiance.

Anthropology approaches normal or paradigmatic science only when its observations and theorizations inculcate material from the older science of linguistics. Levi-Strauss in comparing the differential development of anthropology and linguistics said, "It is true that a discipline whose main, if not sole, aim is to analyze and interpret differences evades all problems when it takes into account only similarities. But at the same time, it thus loses the means of distinguishing between the general truths to which it aspires and the trivialities with which it must be satisfied" (1967, p. 14). And it must be admitted that a certain amount of triviality exists in the preparadigmatic social sciences. The type of concern that usually forms the stock-in-trade of most social scientists makes it conceptually difficult for them to grasp the foundations and implications of structural theory, especially biogenetically oriented structural theory.

Toward a Nomothetic Science of Man

The social sciences, and especially anthropology, are on the verge of becoming normal sciences with a highly coherent body of nomothetic and predictive theory. We suspect this development will soon follow the merger of structural, neurological and evolutionary theory. Although it is inappropriate to discuss the matter here, it is well to mention that the probability of a nomothetic science of man may arouse a good deal of moral trepidation. For what we are suggesting is that the time is rapidly approaching when science will be capable of quite startling and accurate predictions of human behavior.

As a matter of fact, the initial breakthrough we predict is hampered only by the state of present technology. As we have argued above, human behavior is the synthetic result of a dialectic between neurophysiological processes and environmental input. We already have the technological capacity in the development of electronic computers to handle the complex of variables comprising the organism's environment. What we await is the technology by means of which we can trace "holographic" models and connecting tracts in the human central nervous system. Given the capability of tracing and measuring neurophysiological structures and environmental input, accurate prediction of behavior would follow.

Whether or not a future nomothetic science of man will be called "anthropology," "sociology," or whatever is, of course, a moot point, and in a sense irrelevant. No matter what rubric is given to such a paradigmatic discipline, it must be a social science intimately concerned with its biological base, and at least as much concerned with the ways men are alike as with the ways they are different.

References

Altmann, S.A., and C.F. Hockett. 1968. "A Note on Design Features." In *Animal Communication* (ed. by T.A. Sebeok). Bloomington: Indiana University Press.

Beals, R.L., and H. Hoijer. 1971. *An Introduction to Anthropology.* New York: Macmillan.

Braine, M.D.S. 1963. "The Ontogeny of English Phrase Structure: The First Phase." *Langauge,* 39:1-13.

Brown, R., and U. Bellugi. 1964. "The Acquisition of Language." *Monographs of the Society for Research in Child Development,* 29(1).

Brown, R., and C. Fraser. 1963. "The Acquisition of Syntax." In *Verbal Behavior and Learning* (ed. by C.Q. Cofer and B.S. Musgrave). New York: McGraw-Hill.

Chance, M.R.A., and C.J. Jolly. 1970. *Social Groups of Monkeys, Apes and Men.* London: Jonathan Cape.

Gardner, R.A., and B.T. Gardner. 1969. "Teaching Sign Language to a Chimpanzee." *Science,* 165:664-672.

Hallowell, A.I. 1960. "Self, Society and Culture in Phylogenetic Perspective." In

Evolution after Darwin (ed. by Sol Tax), Vol. 11. Chicago: University of Chicago Press.

Harris, M. 1971. *Culture, Man and Nature: An Introduction to General Anthropology*. New York: Thomas Crowell.

Hill, J.H. 1972. "On the Evolutionary Foundations of Language." *American Anthropologist,* 74(3):308-317.

Hockett, C.F. 1960. "The Origin of Speech." *Scientific American* (September).

Isbell, B.J. 1972. "Acquisition of Quechua Morphology." *Papers in Andean Linguistics,* 1(1):79-129. Madison: University of Wisconsin Press.

Jolly, A. 1972. *The Evolution of Primate Behavior.* New York: Macmillan.

Klima, E.S., and U. Bellugi. 1966. "Syntactic Regularities in Speech of Children." In *Psycholinguistic Papers* (ed. by J. Lyons and R. Wales). Edinburgh: Edinburgh University Press.

Kroeber, A.L. 1948. *Anthropology.* New York: Harcourt, Brace.

Kroeber, A.L., and C. Kluckholn. 1963. *Culture: A Critical Review of Concepts and Definitions.* New York: Vintage Books.

Kuhn, T.S. 1970. *The Structure of Scientific Revolutions.* Chicago: University of Chicago Press.

Kummer, H. 1971. *Primate Societies.* New York: Aldine.

Lenneberg, E.H. 1967. *Biological Foundations of Language.* New York: John Wiley.

Levi-Strauss, C. 1967. *Structural Anthropology.* Garden City: Doubleday.

Mann, A. 1972. "Hominid and Cultural Origins." *Man.* 7(3):379-386.

McNeill, D. 1971. "The Capacity for the Ontogenesis of Grammar." In *The Ontogenesis of Grammar* (ed. by D.I. Slobin). New York: Academic Press.

Premack, D. 1971. "Language in Chimpanzee?" *Science,* 172:808-822.

Sade, D.S. 1968. "Inhibition of Son-Mother Mating among Free-Ranging Rhesus Monkeys." *Science and Psychoanalysis,* 12:18-38.

Slobin, D.I. 1966. "The Acquisition of Russian as a Native Language." In *The Genesis of Language* (ed. by F. Smith and G.A. Miller). Cambridge: MIT Press.

——1968. "Early Gramatical Development in Several Languages, with Special Attention to Soviet Research." Working Paper No. 11, Language-Behavior Research Laboratory, University of California, Berkeley.

——1971. "Developmental Psycholinguistics." In *A Survey of Linguistic Science* (ed. by W.O. Dingwall). Published by the Linguistics Program, University of Maryland.

White, L. 1959. *The Evolution of Culture.* New York: McGraw-Hill.

Index